The Golden Stream
A Handbook For The Man Who Keeps Cows For Profit

by Thomas F. Willoughby

with an introduction by Jackson Chambers

This work contains material that was originally published in 1912.

This publication is within the Public Domain.

*This edition is reprinted for educational purposes
and in accordance with all applicable Federal Laws.*

Introduction Copyright 2018 by Jackson Chambers

Self Reliance Books

Get more historic titles on animal and stock breeding, gardening and old fashioned skills by visiting us at:

http://selfreliancebooks.blogspot.com/

Introduction

I am pleased to present another title in the "Cattle" series.

The work is in the Public Domain and is re-printed here in accordance with Federal Laws.

As with all reprinted books of this age that are intended to perfectly reproduce the original edition, considerable pains and effort had to be undertaken to correct fading and sometimes outright damage to existing proofs of this title. At times, this task is quite monumental, requiring an almost total "rebuilding" of some pages from digital proofs of multiple copies. Despite this, imperfections still sometimes exist in the final proof and may detract from the visual appearance of the text.

I hope you enjoy reading this book as much as I enjoyed making it available to readers again.

Jackson Chambers

THE GOLDEN STREAM

THE SELECTION OF GOOD COWS

GOOD cows are the basis of profitable dairying, and the ability to select them is of utmost importance to the dairyman. The use of the Babcock test and a scale is the only certain way to tell just how much milk and butter-fat a cow is producing. But these alone do not tell what she is capable of producing, nor is it possible to use this test in many cases. Because of poor feed and lack of proper care, a cow capable of producing a large profit, would probably not make a very good showing if subjected to a test consisting of weighing and testing her milk. To be of value, such a test must extend through several weeks or months, because one or several milkings do not demonstrate a cow's value. To be profitable, she must be a persistent milker.

Experience and observation show that there is a close relation between conformation and producing power in dairy cows, just as there is in other animals. No man would attempt to make a race horse out of a heavy Percheron, neither would he expect to do heavy draft work with a light trotting horse. Both of these types are very good for certain purposes, but not at all adapted to others. The same is true with cows. The function of the beef-producing animal is to lay on flesh, whereas that of the dairy cow is to produce milk.

There are three qualifications necessary to the good judge of dairy cattle and they are: a knowledge of what kind of conformation makes for the greatest production; a trained eye; and — judgment.

A cow of perfect type is seldom if ever found; therefore, in judging the value of a cow a man must understand the relative importance of different features of her conformation. He must know that a finely shaped udder does not always make a good cow; that a pretty head or a skin of fine texture do not overcome the disadvantages of a weak constitution or lack of capacity to consume enough feed to produce a large flow of milk.

In teaching the judging of dairy cattle and other livestock, score cards are used on which 100 represents the standard of perfection. These score cards are very helpful, for the purpose of developing a systematic way of picking out the good and bad features of an animal and to make it easier for the beginner to locate the different points to be considered. After the eye becomes trained and the ideal type becomes fixed in the mind, the score card can be discarded. The score card commonly used, and a picture of an excellent type of dairy cow with the different features of her conformation pointed out, will be found on subsequent pages of this book.

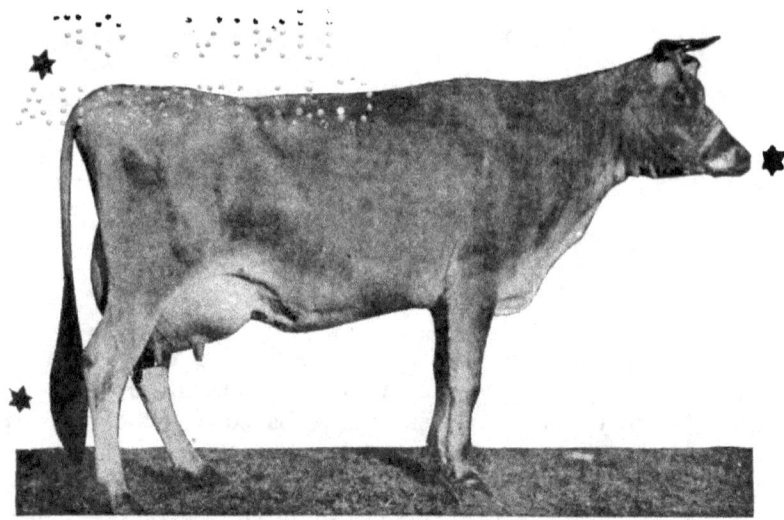

Side View of a Dairy Cow. Wedge shape indicated by stars

In judging a cow it is well to go about the work systematically, and consider her in sections. These are classed under the heads of general appearance, head, forequarters, body, hindquarters, and mammary development.

General Appearance — Fifteen points out of 100 are given to general appearance. Viewed from the sides and front or rear, the outlines of the cow are similar to those of wedges. The bases and points of these wedges are indicated in accompanying illustrations.

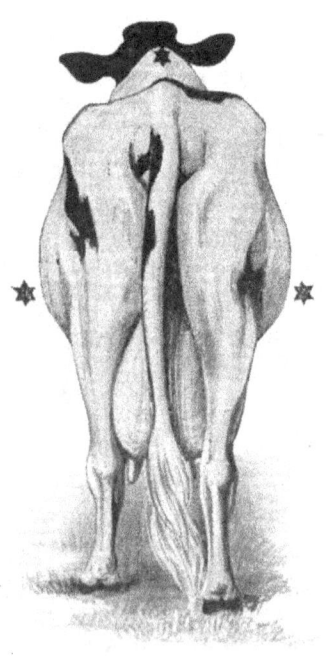

Rear View of a Dairy Cow. Wedge shape indicated by stars

The dairy cow should be thin, angular, and loose-jointed. At the same time she should have the appearance of strength and vitality. She should be thin because she is turning a large part of the feed she eats into milk, not because she is poorly fed or diseased. In disposition she should be quiet, yet keenly alive to what is going on about her. Her skin should be soft and pliable with an abundance of secretion, and the hair fine.

Head — The head should be clean cut and refined. The good muzzle is large with large, open nostrils. A small narrow muzzle indicates a weak constitution and a lack of capacity for consuming larger quantities of feed. The face extends from the muzzle to the forehead and should be of medium length and clean cut. Such a face indicates refinement and dairy temperament. A common defect is a long, "horsey" face, which is generally accompanied by a narrow forehead. The forehead should be broad and slightly dished. A broad forehead indicates the well-developed, nervous system found in the heavy milker.

Feel the lower jaw and see that it is clean cut, strong, and firmly attached to the upper jaw. Large, bright eyes indicate intelligence.

Forequarters—The neck should be thin and free from loose, flabby skin. A coarse, beefy neck and throat indicate lack of dairy character and should be discriminated against. The withers should be thin and sharp. The shoulders should slope outward, giving a large chest capacity. The fore legs should be straight and clean cut. They should be set well apart to allow ample chest capacity.

Body—In this part of the animal are located the vital organs and the digestive organs, consequently the conformation here is very important. Capacity to consume large quantities of feed is indicated by a large, deep barrel. The ribs should be well sprung. A common defect is that the ribs do not spring out enough as they extend down. The result is that, while fairly deep, the barrel is narrow and consequently lacking in capacity. In passing the hand over the ribs they should be found set well apart; the last two or three ribs should be sufficiently far apart to permit two or three fingers to be inserted between them.

In examining the back, stand off a pace or two and note whether it is straight or not. A slight sagging in the back is often found and, while this should not be discriminated against too severely, it is an indication of weakness. Pass the hand along the spine and see that the vertebrae are not closely joined. The lateral nerves from the spinal cord pass out between the vertebrae and ample space for this is necessary, as nervous development is very important to the dairy cow. The loin should be broad, long, and have the appearance of strength. Weak, narrow loins are very common.

A deep, broad, floored chest is one of the best indications of constitution. The depth can be noted by standing back a few paces from the cow; but in judging the width of the chest, pass the hand under the body just back of the forelegs. In many cases it will be found that cows which have a deep chest are narrow in this part; such conformation should be discriminated against, as depth alone is not sufficient to insure ample room for the heart and lungs.

Hindquarters—The hips should be prominent and wide apart. The rump should be high, long and carry well out behind. A common defect is a rump that slopes down. The

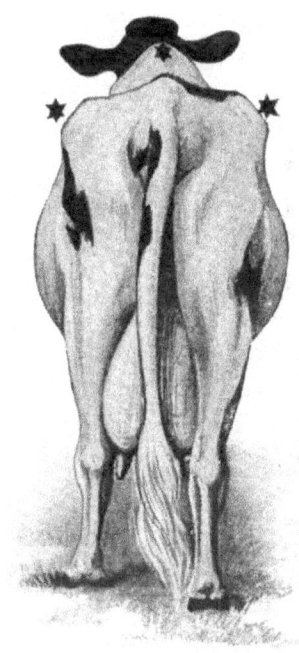

Wedge shape indicated by stars as seen when looking along the animal's back

A Row of Money Makers

position of the pin bones and the rump are closely related. The rump should be well arched and the pin bones set well apart, as this is a great help towards easy parturition. The tail is a good indication of quality. A long, thin, tapering tail is desirable, while a thick, coarse tail is an indication of coarseness throughout the animal. The hind legs should be trim and set well apart. Hind legs that are crowded close together are objectionable, as they limit the space for the udder. The thighs should be thin and practically all the width and flesh carried well up. Thick, chunky thighs are to be discriminated against, as they indicate a tendency to beefiness and reduce the space for udder development.

Mammary Development—A large, well-shaped, well-placed udder is of the utmost importance. This part of the cow should be very carefully examined, as a large part of the value of the cow will be determined by the formation of her udder and milk veins. On the score card 30 points of a possible 100 are given to the mammary development.

Look at the udder from the sides and the rear and note its general shape and the manner of attachment in front and in rear. The floor or bottom of the udder should be straight and extend well forward. It should also extend well back. It should be attached well up in the rear. The quarters should be balanced and the teats squarely placed.

While size is very important, a careful examination should be made to see that the size is not due to meatiness or coarseness. A cow with a large, coarse, meaty udder is, as a rule, neither a heavy nor a persistent milker. The udder should be soft and of a very fine, spongy texture. When milked out, it should collapse into numerous folds. Examine the teats and see that they are evenly placed and of medium and uniform size. Milk out a few streams and see that the openings are free from obstructions. Too small openings are undesirable, as they make the cow hard to milk.

The milk veins, which extend from the udder forward and pass into the body through openings in the body wall known as milk wells, indicate the amount of blood that passes through the udder. The supply of blood to the udder determines the flow of milk the udder will secrete, and therefore an examination of the milk veins is important. They should be large, tortuous, extend well forward along the cow's belly, and have numerous branches. The milk wells or openings through which the veins pass into the body should be as numerous as the veins and branches.

After the details of the animal's conformation have been carefully gone over, the judge should step back a few paces and walk around the animal several times, and in this way review the points he has gone over and weigh the value or lack of value of the good features or defects discovered in the examination.

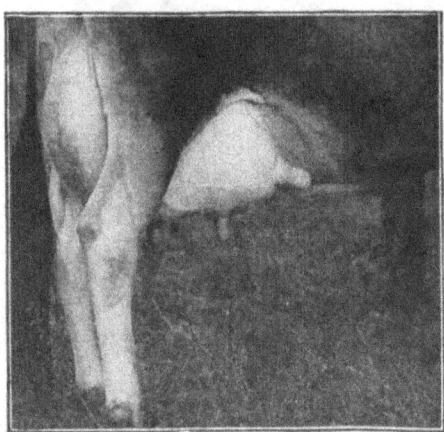

A large, well shaped, well placed udder

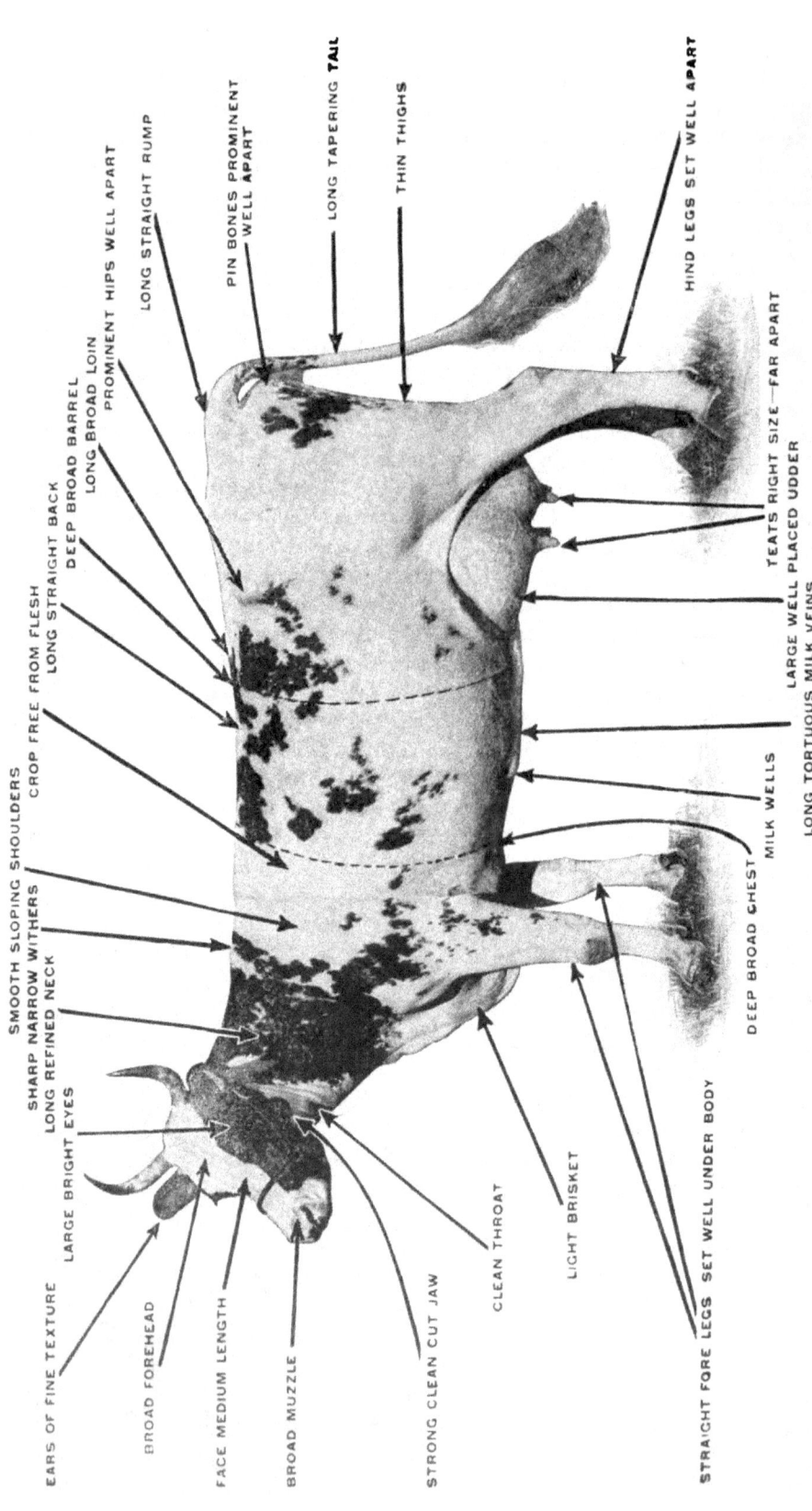

A Cow of excellent dairy type, with points to be considered in judging a cow indicated

THE SCORE CARD FOR JUDGING DAIRY COWS

SCALE OF POINTS	Standard
HEAD—8 Points	
1. Muzzle, broad	1
2. Jaw, strong, firmly joined	1
3. Face, medium length, clean	1
4. Forehead, broad between eye dishing	1
5. Eyes, large, full, mild and bright	2
6. Ears, medium size, fine texture, secretions oily and abundant, yellow color	2
FOREQUARTERS—10 Points	
7. Throat, clean	1
8. Neck, long, spare, smoothly joined to shoulders, free from dewlap	2
9. Withers, narrow, sharp	3
10. Shoulders, sloping, smooth; brisket, light	3
11. Fore Legs, straight, clean, well set under body	1
BODY—25 Points	
12. Crops, free from fleshiness	1
13. Chest, deep, roomy; floor broad	6
14. Back, straight, strong; vertebræ open	3
15. Ribs, long, deep and well sprung	3
16. Barrel, deep, long, capacious	10
17. Loin, broad, strong	2
HINDQUARTERS—12 Points	
18. Hips, prominent, wide apart	1
19. Rump, long, level, not sloping	4
20. Pin Bones, wide apart	1
21. Tail, neatly set on, long, tapering	1
22. Thighs, spare, not fleshy	3
23. Hind Legs, well apart, giving ample room for udder	2
MAMMARY DEVELOPMENT—30 Points	
24. Udder, large, very flexible, attached high behind, carrying well forward; quarters even	15
25. Teats, wide apart, uniformly placed, convenient size	5
26. Milk Veins, large, tortuous, extending well forward	4
27. Milk wells, large	6
GENERAL APPEARANCE—15 Points	
28. Disposition, quiet, gentle	2
29. Health, thrifty, vigorous	3
30. Quality, free from coarseness throughout; skin soft, pliable; secretions abundant; hair fine	4
31. Temperament, inherent tendency to dairy performance	6
Total	100

Champion Jersey Herd, National Dairy Show, 1911

Champion Jersey Cow, National Dairy Show, 1911

WHERE THE JERSEY ORIGINATED

Jersey cattle are famous the world over for their rich milk. They were the first dairy breed to attract public attention to any extent in this country. This breed has the following very important characteristics:

1. They convert a large part of the food consumed into milk and not into flesh and fat.

2. They give the richest milk.

3. They mature at an early age; hence they can be bred early, thus avoiding the necessity of waiting long periods before they come into usefulness.

Jerseys are the most famous of the Channel Island breeds and they originated on the Island of Jersey, which is the largest and most important of the Channel Islands. This island is only about eleven miles from east to west and averages about five and one-half miles in width. The land is rich and very productive. On account of the Norman law of succession, Jersey farms have become very much subdivided, and it is only occasionally that they exceed fifty acres, while many are less than three acres. The farm houses and cottages are remarkably neat and comfortable, and the people, who all farm their own land, are perhaps the most contented and prosperous in the United Kingdom. The pasturage is very rich and is much improved by the application of sea weed to the surface. The mainstay of the island is its cattle, and this breed is kept pure by stringent laws against the importation of foreign animals. The milk is used almost exclusively to manufacture butter.

Champion Holstein Herd, Iowa Dairy Show, 1911

Champion Holstein Cow, National Dairy Show, 1911

THE HOLSTEIN-FRIESIAN CATTLE

The Holstein cattle have been in existence as a breed of dairy cattle for over 2,000 years. A people known as Friesians, who came presumably from the shores of the Baltic, settled about the year 300 B. C. in the valley of the Rhine, Germany. These people brought with them their white cattle. One hundred years later, another tribe called the Batavians, came to this same territory along the Rhine with their herds of black cattle. The combination of these two herds produced the black and white breeds of Europe.

These cattle were introduced into America about the year 1625 by the early Dutch settlers. Further importations were made in 1810. These early animals were probably bred to native cattle, with the result that the purity of the blood was lost. The first cow to which we can directly trace any of this breed was imported to the United States in 1852.

The first cattle of this breed were given the name Holstein by the importer. The name Friesian was given the breed by another and later importer, who called his cattle Dutch Friesians. The name was later changed to Holstein Friesians.

The true type of this breed is the result of centuries of selection and environment. The breed is noted for marvelous milk production, powerful digestion, and perfect assimilation of food.

Instances have been recorded where a cow of this breed produced in one year as much as 30,000 pounds of milk, and there are many records over 20,000 pounds. However, their milk is not as rich in butter fat as that of the Jersey or Guernsey. The large size of the Holstein is the first thing to impress the casual observer. Next to the Jersey, the Holsteins are second in point of numbers in the United States.

Champion Guernsey Herd, Iowa Dairy Show, Waterloo, Ia., 1911

'Champion Guernsey Cow,' National Dairy Show, 1911

A FEW FACTS ABOUT GUERNSEYS

Guernseys, like Jerseys, are a Channel Island breed, having originated on the Island of Guernsey in the English Channel. These breeds doubtless had a common origin, although they are at present bred and developed independently. Both have been developed as dairy cattle, and they resemble each other in general appearance and in characteristics. Guernsey cattle are somewhat larger than Jerseys, also coarser in bone and carry more flesh. They are noted for the rich, yellow color of their milk and cream. Next to the Jersey, the Guernsey produces the richest milk. In quantity, the Guernsey yield often excels that of the Jersey.

In America in the early days, the Jerseys and Guernseys were classed together under the general name of Alderney, but later they were recognized as separate breeds.

The Guernseys are prolific milkers, and their gentle disposition, much like that of the Jersey, makes the breed a favorite. The Guernsey is an ideal family cow, as it is a light feeder, but rich in milk production. Five thousand pounds of milk and over in a year is not at all an unusual performance for a Guernsey.

The Guernsey has a finely shaped head, a long, slender neck, large and deep body conformation, and thin, shapely flanks. The color is light yellow, reddish, or fawn, with white spots on the legs and body.

The Guernseys have become very popular in America, because they have strong constitutions, are good feeders, and produce a large flow of rich milk. One of the most famous cows of this breed is Yeksa Sunbeam, who produced in one year 14,920.8 pounds of milk, containing 857.15 pounds of butter. The average test of this milk was 5.74.

Champion Ayrshire Herd, Iowa Dairy Show, 1911

Champion Ayrshire Cow, National Dairy Show, 1911

THE AYRSHIRE CATTLE

This well-known breed was originated in the mountainous county of Ayrshire, located in southwestern Scotland, and brought to its present standard by careful breeding in this country.

The Ayrshires first appeared in this country in the State of New York in the early part of the nineteenth century, and their numbers were considerably increased about the middle of the century.

The Ayrshire breed is famous for its economy in feeding and the ease with which it withstands conditions that would be a serious hindrance to other breeds in the production of milk.

The Ayrshire is of a nervous disposition and is apt to be quarrelsome at times. The markings of the Ayrshire are red and white in spots, not mixed, with a tendency at present toward more white. In size the Ayrshire is about the size of the Dutch Belted type.

The Ayrshire cow weighs from 900 to 1,100 pounds. Records show that individual cows have produced as high as 10,000 and 12,000 pounds of milk a year. Butter fat in the milk averages about 4 per cent. Because of its composition, Ayrshire milk is especially well adapted for shipment to city markets.

This breed has become very popular in America within the last few years, and in this respect they are a close rival of the Jerseys, Guernseys and Holsteins. One of the greatest triumphs of this breed came when Oldhall Lady Smith 4th won the Grand Championship over all breeds at the 1911 National Dairy Show.

The owners of Ayrshire cattle in America have as a rule been practical dairymen who have not forced their cows in attempts to make phenomenal records. The breed as a whole, therefore, is in a much more normal condition than some of the other prominent dairy breeds.

A Prize Winning Brown Swiss Herd

A representative Brown Swiss Cow

THE BROWN SWISS CATTLE

Switzerland has long been famous as a dairy country. The Brown Swiss cattle of that country, grazed on Alpine grass, are remarkably strong and healthy. Their native home is the canton of Schwyz, which is the most noted canton for the dairy industry.

The first importation of these cattle into America was made by Henry M. Clarke, of Belmont, Mass., in 1869. Many other importations followed and there are somewhat over 5,000 of these cattle in the New England and the Middle and Western States.

The color of these cattle is usually described as being brown. The color, however, runs through many shades and is often a mouse color or brownish dun. The darkest color is on the head, neck, and legs. The nose, tongue, hoofs, and switch are black. The average weight of the cows is 1,300 or 1,400 pounds, and the bulls weigh from 1,500 to 2,500 pounds.

There are numerous instances where cows of this breed have produced as high as 10,000 pounds of milk in a year and in some cases a production of 16,000 pounds has been reached. The milk from these cows has an average test of about 4 per cent butter fat.

Long life is one of the pronounced characteristics of this breed. The cows are in their prime when eleven and twelve years old and frequently continue to breed up to sixteen or eighteen years of age. They are strong and of a somewhat stolid disposition. While it is true that the cows of this breed do not conform as closely as do those of some other breeds to the recognized dairy type, they prove persistent and profitable producers. In conformation they have a tendency to be round, plump and compact.

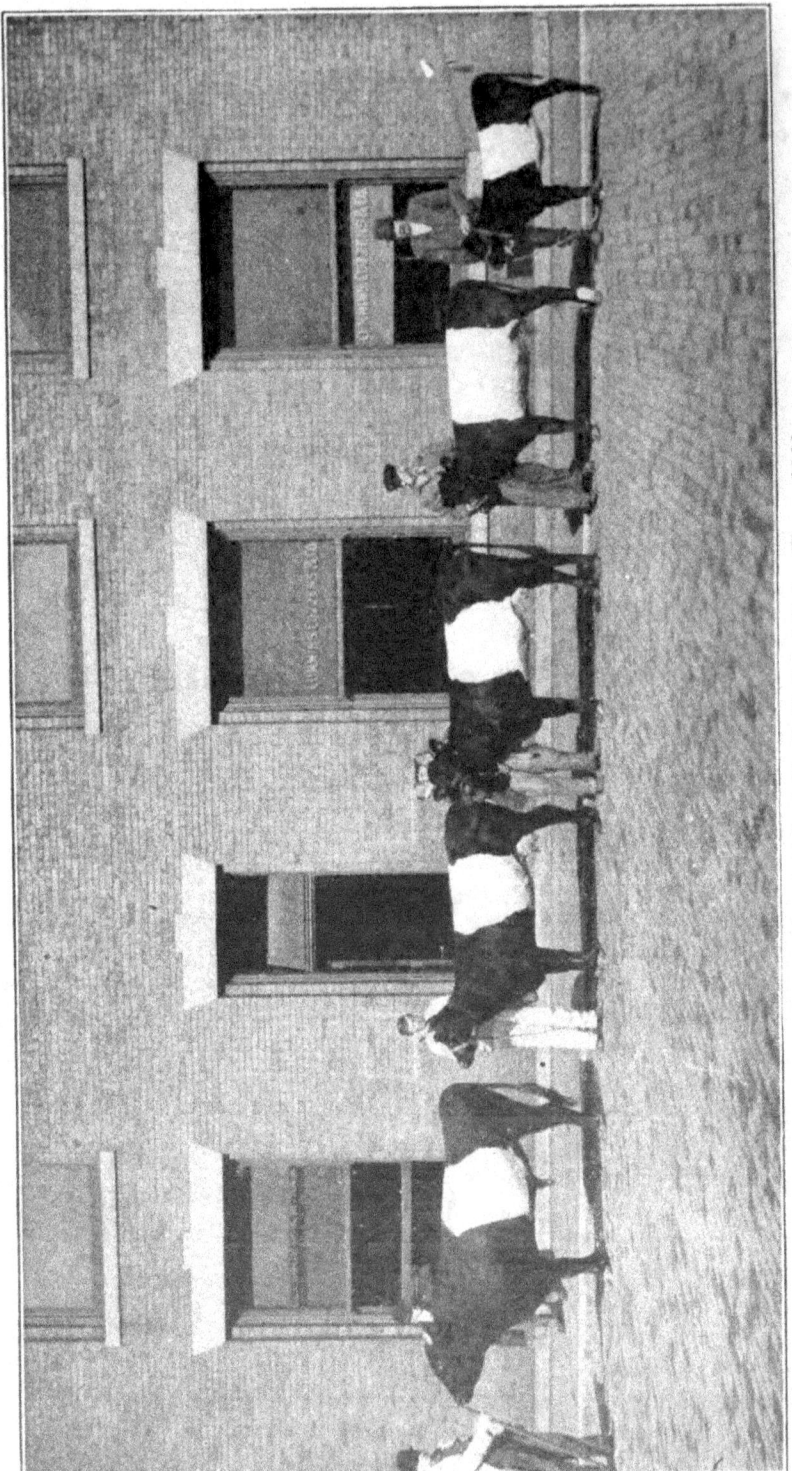

Champion Dutch Belted Herd, National Dairy Show, 1911

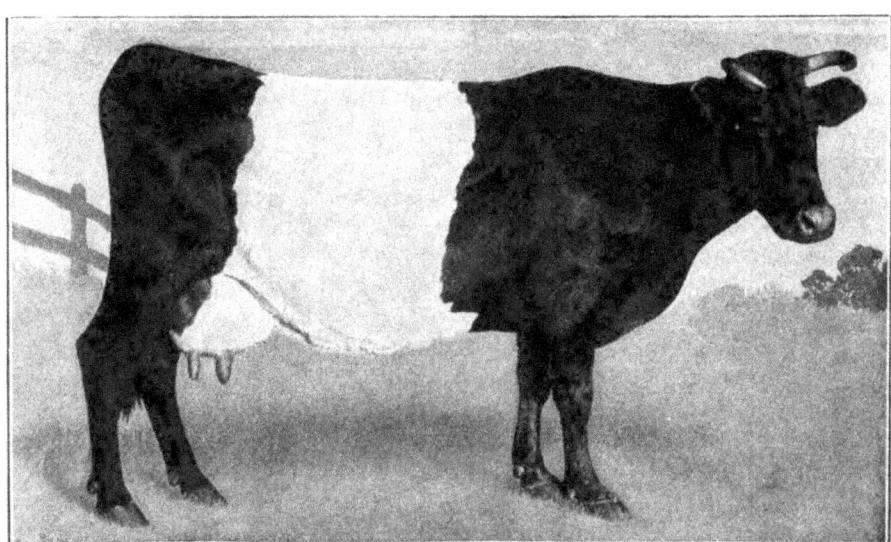

An excellent type of Dutch Belted Cow

THE DUTCH BELTED CATTLE

The history of Dutch Belted cattle indicates that these cattle first began to flourish in considerable numbers in 1750. This breed is distinguished by a white belt around the body, the balance of the body being black. In Holland the breed is known as Lakenvelders, which literally means a field of white, or in this case, a white body with black ends.

Regarding the history of this breed in America, Bailey's Cyclopedia of American Agriculture says: "Dutch Belted cattle were first imported to America in 1838. D. H. Haight was the largest importer. He made his first importation in 1838, and a later one in 1848. His herd became scattered over Orange County, N. Y., until one will find a great many belted cows in every township in that county today. Hon. Robert W. Coleman also imported a large herd to place on his estate at Cornwall, Pa. The Dutch Belted cattle in America today are entirely descended from these herds. In 1840, P. T. Barnum imported a number of Dutch Belted cattle for show purposes, but shortly placed them on his farm in Orange County, N. Y. One heifer was imported in 1906 by Dr. H. W. Lane, of New York City, for his farm in New Jersey, but previous to that time none were brought over for more than fifty years. This was due chiefly to the very great difficulty in securing them and to the restrictions against importing them. A number have been exported from this country to Canada and Mexico, and a few to Cuba. In 1893, H. B. Richards, Secretary of the Dutch Belted Cattle Association, sold his World's Fair herd, numbering sixteen, and nine others to a son-in-law of President Diaz, and shipped them to Mexico. Later, Mr. Richards sold twenty to Sir William Van Horne, of Canada. Other exportations have been made. There are about fifteen hundred head in America at the present time."

The cows weigh on an average of 1,000 to 1,200 pounds. The Dutch Belted is very similar to the Holstein in the amount and quality of the milk produced.

PROFITABLE FEEDING

Next to the feeding of "scrub" cows, the "scrub" feeding of good cows is responsible for the small profits many dairymen realize. No cow can be expected to produce the large flow of milk that she must produce to be profitable, unless she receives a sufficient quantity of feed of the right quality. To profitably feed dairy cows the feeder must have a knowledge of what the cow requires and of what kind of feed will best meet these requirements. There are no short cuts to this knowledge. It is secured only by spending time and energy in the study of the subject. Efforts in this direction are, however, well worth while. Every farmer who has conscientiously endeavored to learn what feed his cows require and has then supplied it, has been amply paid for his trouble in the increased profits secured.

Much has been said about feeding balanced rations, but in spite of this the balanced ration is very much misunderstood. Many farmers imagine that it is something new-fangled and impractical. As a matter of fact, it is simply feeding a cow, in as exact a manner as is possible, the feed she needs to sustain life and manufacture a large quantity of milk. Chemical analysis of our common feeding stuffs show that they contain all the necessary material for body maintenance and milk production, but in practically no instances are these materials contained in any one kind of feed in the right proportion for the most satisfactory results. Feeding a balanced ration, then, is simply feeding a ration in which the nutrients contained are in the right quantity and proportion to secure maximum results at a minimum cost.

The three materials, or nutrients, contained in feeding stuffs that must be considered by the feeder of dairy cows are: protein, carbohydrate and fat. To take these up separately, we find:

First, protein is the most expensive and difficult material to secure. Ordinary farm-grown feeds are, as a rule, deficient in protein. In the animal body this material goes to build up muscle, blood and connective tissue. In milk it is represented by casein and albumen. Among farm crops it is found in the largest quantities in leguminous plants, such as alfalfa, clover, cow peas and soy beans. Most dairymen secure a considerable part of the protein they feed through the purchase of highly concentrated feeds, such as linseed meal, cottonseed meal, gluten feed, bran, etc. Protein should compose about one-sixth of the nutrients in the ration of the average cow when in milk.

Second, carbohydrate is that part of the feed which goes to produce bodily heat, energy and fat. In milk, it is found in the form of sugar and fat. The problem of securing sufficient carbohydrate is never a serious one, as it is the cheapest and most abundant and is found in large quantities in most farm crops.

Third, the fat or oil is a nutrient that for all practical purposes should be considered the same as carbohydrate. It fills the same purpose as carbohydrate and contains the same elements, but in a more concentrated form. Experiments show that one pound of fat is equal to 2.25 pounds of carbohydrate. In the calculation of rations, carbohydrate and fat are considered together. That is, the fat is multiplied by 2.25 and then added to the carbohydrate.

In making up rations for dairy cows, those feeds which are grown on the farm should be used as extensively as possible. Feeds which are grown on the farm are much cheaper than those which must be purchased, and practically the only feeds that the average farmer needs to buy are those rich in protein.

Corn silage should always be a part of the dairy cow's ration, with the exception of the time when the cow is on rich pasture. Corn silage, however, is not a balanced ration. Some grain and hay should be fed with it. From thirty to forty pounds of silage a day, fed in two feeds, will be sufficient for a cow unless she is a very large animal, in which case the quantity can be slightly increased.

One of the greatest mistakes that many dairymen make is that of feeding timothy hay to milk cows. Timothy hay has its uses, but much better feeds can be found for milk-producing cows. Alfalfa, clover, cowpea hay, vetch hay, soybean hay, and velvet bean hay are crops, one or more of which are adapted to most localities, which furnish the most desirable dry roughage for dairy cows.

Local prices, to some extent, must be taken into consideration when selecting the concentrated or grain portion of the ration. The price of the feeds varies in different localities. Hence, in buying concentrates two things should be considered: First, and above all else, the nutritive value of the feed. Second, prices on the local market.

On the following page the method of formulating a balanced ration is explained and a feeding standard for dairy cows given. It should be remembered, however, that no two cows are exactly alike; and therefore the kind and quantity of feed will vary with individuals. The feeder, then, must not only be familiar with the theoretical requirements, but must make a study of the individual requirements of each cow in his herd.

AMOUNT OF FEED REQUIRED BY DAIRY COWS

The following feeding standards are the results of investigations, by Haecker, into the requirements of dairy cows and will serve as a very good guide to the dairyman in mixing rations for his cows that will meet the requirements of bodily maintenance and maximum milk yield.

For every 100 pounds of live weight there is required a maintenance allowance of .07 of a pound of protein, 0.7 of a pound of carbohydrate, and .01 of a pound of fat.

The following is Haecker's feeding standard for dairy cows. To the quantity of nutrients given below should be added the allowance for maintenance given above. The 1,000-pound cow is taken as a basis in the following table:

	Daily Allowance of Digestible Nutrients		
	Crude Protein Lbs.	Carbohydrate Lbs.	Fat Lbs.
For each pound of 3.0 per cent milk	0.040	0.19	0.015
For each pound of 3.5 per cent milk	0.042	0.21	0.016
For each pound of 4.0 per cent milk	0.047	0.23	0.018
For each pound of 4.5 per cent milk	0.049	0.26	0.020
For each pound of 5.0 per cent milk	0.051	0.27	0.021
For each pound of 5.5 per cent milk	0.054	0.29	0.022
For each pound of 6.0 per cent milk	0.057	0.31	0.024
For each pound of 6.5 per cent milk	0.061	0.33	0.025
For each pound of 7.0 per cent milk	0.063	0.35	0.027

To illustrate how this table is used, let us assume that we have a cow weighing 1,200 pounds and producing thirty-five pounds of 4 per cent milk a day. Multiplying the maintenance allowance for 100 pounds of live weight by 12 the results show that a 1,200-pound cow requires per day for maintenance .84 of a pound of crude protein, 8.4 pounds of carbohydrate, and .12 of a pound of fat. For the production of thirty-five pounds of 4 per cent milk she requires thirty-five times the allowance given in the table for the production of one pound of 4 per cent milk. We find that to produce this milk she must receive above the maintenance allowance 1.545 pounds of protein, 8.05 pounds of carbohydrates and .63 of a pound of fat. Tabulating these results and adding the maintenance and producing allowance, we have the following results:

	Crude Protein Lbs.	Carbohydrate Lbs.	Fat Lbs.
For Maintenance	.84	8.4	.12
For producing 35 pounds of 4 per cent milk	1.545	8.05	.63
Total nutrients required per day	2.385	16.45	.75

In formulating rations for dairy cows the roughage portion of the ration should be taken as a basis and its deficiencies overcome by the addition of concentrates. Take, for instance, the 1,200-pound cow producing thirty-five pounds of 4 per cent milk a day. This cow should have about forty-five pounds of corn silage and ten pounds of hay, and for the sake of illustrating the formulation of a ration, we will use clover hay. By referring to the table on page 27 we find that the average nutrients contained in 100 pounds of corn silage are: Protein, 1.4 pounds; carbohydrates, 14.2 pounds, and fat, .7 of a pound The nutrients in forty-five pounds of silage are found as follows:

Protein, $1.4 \div 100 \times 45 = .63$ lbs. of protein in 45 lbs. of silage.

Carbohydrates, $14.2 \div 100 \times 45 = 7.57$ lbs. of carbohydrates in 45 lbs. of silage.

Fat, $.7 \div 100 \times 45 = .311$ lbs of fat in 45 lbs. of silage.

The table also shows that 100 pounds of clover hay contains 7.1 pound of protein, 37.8 pounds of carbohydrate and 1.8 pounds of fat. The amount of these nutrients in ten pounds of clover (red) hay are, therefore:

Protein, $7.1 \div 100 \times 10 = .71$ lbs. of protein in 10 lbs. of clover hay.

Carbohydrates, $37.8 \div 100 \times 10 = 3.78$ lbs. of carbohydrates in 10 lbs. of clover hay.

Fat, $1.8 \div 100 \times 10 = .18$ lbs. of fat in 10 lbs. of clover hay.

Adding the nutrients in forty-five pounds of silage and in ten pounds of clover hay and subtracting their sum from the total nutrients required by a 1,200-pound cow producing thirty-five pounds of 4 per cent milk, we find the nutrients which must be supplied by the concentrated portion of the ration.

	Crude Protein Lbs.	Carbohydrate Lbs.	Fat Lbs.
Total nutrients required per day by a 1,200-lb. cow	2.385	16.45	.75
Total nutrients supplied by 45 pounds of silage and 10 pounds of clover hay	1.34	11.35	.495
Pounds of nutrients which must be supplied by the concentrated portion of the ration	1.045	5.10	.255

In feeding the concentrated portion of the ration, it will be found very convenient and practical to make up several hundred pounds of a mixture of several concentrates and then feed such quantities of this mixture as are necessary to make up the deficiency of the roughage portion of the ration. For the purpose of illustration, we will assume that the concentrates at hand for feeding to the above mentioned cow are: Corn meal, wheat bran and gluten meal.

For a trial ration we will take four pounds of corn meal, four pounds of wheat bran and two pounds of gluten meal. Using the same process as was used in finding the nutrients contained in the roughage portion of the ration, we find:

	Crude Protein Lbs.	Carbohydrate Lbs.	Fat Lbs.
4 pounds of corn meal	.244	2.572	.14
4 pounds of wheat bran	.476	1.68	.1
2 pounds of gluten meal	.594	.85	.122
Total	1.314	5.102	.362

Adding to this the nutrients contained in forty-five pounds of silage and ten pounds of hay, we have 2.654 pounds of protein, 16.452 pounds of carbohydrates and .857 of a pound of fat.

Comparing these results with the standard, it will be noted that there is a slight excess of protein and of fat. This variation from the theoretical standard is, however, permissible. It should be remembered that if commercial feeding stuffs fall below standard, the probability is that the deficiency will be in protein. It is also practically impossible to formulate a ration that will exactly meet the requirements of the standard, but every effort should be made to adhere as closely as possible to the standard.

Our ration, then, for a 1,200-pound cow, giving thirty-five pounds of 4 per cent milk, is: forty-five pounds corn silage, ten pounds clover hay, four pounds corn meal, four pounds wheat bran and two pounds gluten meal.

There are many different kinds of feed that can be used in formulating a ration for a dairy cow and it should be understood that the above feeds are used simply as an illustration.

The table on the opposite page gives the amount of nutrients contained in a number of the more common feeding stuffs. These analyses are taken from Henry's "Feeds and Feeding:"

Jacoba Irene, a Jersey cow who produced in two years (December 11, 1906, to January 24, 1909—dry 45 days) 31,508.9 pounds of milk, containing 1,745.06 pounds of butter fat

KIND OF FEED	Dry matter in 100 lbs.	Digestible Nutrients		
		Crude Protein	Carbo-hydrate	Fat
ROUGHAGE				
Corn fodder with ears on	57.8	2.5	34.6	1.2
Corn fodder, ears removed	59.5	1.4	31.2	0.7
Corn silage	26.4	1.4	14.2	0.7
Hay from mixed grasses	84.7	4.2	42.0	1.3
Kentucky blue grass	86.0	4.4	40.2	0.7
Red clover	84.7	7.1	37.8	1.8
Soybean hay	88.2	10.6	40.9	1.2
Cowpea hay	89.5	9.2	39.3	1.3
Alfalfa	91.9	10.5	40.5	0.9
Hairy vetch	88.7	11.9	40.7	1.6
Peanut vine	92.4	6.7	42.2	3.0
Velvet bean	90.0	9.6	52.5	1.4
Mixed grasses and clover	87.1	5.8	41.8	1.8
CONCENTRATES				
Dent corn	89.4	7.8	66.8	4.3
Flint corn	88.7	8.0	66.2	4.3
Corn meal	85.0	6.1	64.3	3.5
Corn and cob meal	84.9	4.4	60.0	2.9
Gluten feed	90.8	21.3	52.8	2.9
Gluten meal	90.5	29.7	43.5	6.1
Standard wheat middlings (shorts)	88.8	13.0	45.7	4.5
Wheat bran	88.1	11.9	42.0	2.5
Wheat screenings	88.4	9.6	48.2	1.9
Rye	91.8	9.5	69.4	1.2
Barley	89.2	8.4	65.3	1.6
Oats	89.6	8.8	49.2	4.3
Ground oats	88.0	10.1	52.5	3.7
Oat middlings	91.2	13.1	57.7	6.5
Cowpea	85.4	16.8	54.9	1.1
Soybean	88.8	29.1	28.3	14.6
Kafir corn	90.1	5.2	44.3	1.4
Linseed meal (old process)	90.2	30.2	32.0	6.9
Linseed meal (new process)	91.0	31.5	35.7	2.4
Cotton seed	89.7	12.5	30.0	17.3
Cotton seed meal	93.0	37.6	21.4	9.6
Cotton seed hulls	88.9	0.3	33.2	1.7
Dried brewers' grains	91.3	20.0	32.2	6.0
Dried distillers' grains	92.4	32.8	39.7	11.6
Dried beet pulp	91.6	4.1	64.9	
Sugar beet molasses	79.2	4.7	54.1	
Alfalmo	90.9	9.8	40.8	0.9

DEVELOPING THE DAIRY HEIFER

More and more, thoughtful dairymen are coming to realize that in order to have profitable herds they must raise and develop their own cows. Men who have good cows are not anxious to part with them, and the dairyman who depends upon the purchase of mature cows to keep up his herd is forced to pay someone a handsome profit for raising them, or else be content with the culls of other herds. Even when he pays the high price he is not certain that the cows he buys will prove a profitable investment.

It will be found much cheaper and more certain for the average dairyman to raise his own cows than to pay someone a profit for raising them for him. It is a fact that most of the best-producing herds in the country have been bred and raised by the men who own them.

The heifer calves raised should be those from the best cows, and at least one-sixth as many heifers should be raised as there are cows in the herd. It is estimated that on an average, one-sixth of the cows in a herd each year reach an age when it is no longer profitable to keep them. Therefore, by raising one-sixth as many heifer calves as there are cows in the herd, those cows which must be disposed of from year to year will be replaced.

The heifer calf which is to be raised for a future producer in the herd should, of course, be the offspring of animals of proved milk-producing power. But that will not necessarily mean that she will develop into a good cow. The care and feed the calf receives from birth to maturity is of utmost importance. The best heifer calf in the world can easily be ruined as far as future milk-production is concerned, by improper feeding and treatment. There are a few simple rules, which, if followed and

It pays to give calves good feed and care

supplemented by a litttle judgment and experience. will produce results well worth the effort.

During the first three or four days the calf may be permitted to run with the cow, or it may be taken away after it is a few hours old. Both methods are used successfully. For at least two, and preferably three, weeks, the calf should be fed its mother's milk out of a clean pail three or four times a day. The number of feedings will depend upon the strength of the calf, and the amount will also have to be determined in a like manner. An average calf, however, should receive four to six pounds of its mother's milk at a feeding for the first two or three weeks. The change from whole milk to skim milk should be gradual, and by the time the calf is on a skim-milk diet the number of feedings should be cut down to two a day.

The amount of skim milk a calf will consume is not an indication of the amount to feed. A calf will drink more skim milk than is good for it. Feed so that the calf will look for more when the pail is empty, but don't give more. By the time the calf is three months old it should be receiving about twenty pounds of skim milk a day, the increase to this amount being gradual. Don't try to force growth by heavy feeding or the result will be a sickly, stunted calf.

An average calf should, however, receive about twelve pounds of skim milk a day until it is six weeks old. This should be gradually increased so that the calf is drinking at least twenty pounds a day by the time it is three months old. By this time the calf will be eating enough food other than skim milk so that it will not be necessary to increase the milk ration over twenty pounds per day.

A great deal of the success of calf-feeding depends upon the judgment of the person who is doing the feeding. This judgment is not an accidental acquisition, but instead is the result of careful study.

One of the most vital considerations in feeding the calf is to have the milk—whether whole or skim milk—warm and sweet, and just as fresh from the cow as is possible. It is here that the hand separator is valuable. While separator skim milk does not contain as much fat as gravity skim milk, it is clean, warm and wholesome—which often is not the case with gravity skim milk. The pails from which the milk is fed should be frequently cleaned and scalded, so that the milk which the calf drinks will not be contaminated. If clean, warm skim milk is regularly fed from pails that are kept in a sanitary condition, and the amount of milk and time of feeding are properly regulated, there is practically no danger of scours.

A calf will not do well on skim milk alone, and consequently it should be taught to eat a little grain as soon as possible. Many feeders add a little oil meal to the skim milk and let the calf lick it out of the bottom of the pail, but as the calf must learn sooner or later to eat grain without milk, there is nothing gained by postponing teaching the calf to eat in the proper way. After the calf has finished its milk do not let it out of the stanchion, but instead put a little grain before it and it will soon nose around and begin to eat. A good way to get the calf interested in the grain is to put a handful in its mouth.

A mixture of equal parts of bran and ground oats forms an excellent ration for the dairy calf. Incidentally, it may be well to mention the fact that oats (unground) are one of the best cures for scours. In case the calf is very thin, a little corn meal can be added to the ration.

Not only should the dairy calf have a ration of grain, but it should also be encouraged to eat a large amount of roughage, as this will have a tendency to develop a capacity for consuming a large amount of food. Capacity is one of the most important characteristics of a good cow and every effort should be made to develop it.

If it is possible give the calf an abundance of alfalfa hay, as it is one of the best growth-producing feeds in the world, and besides it has a very good effect on the digestive system of the calf. When alfalfa is fed there is practically no need of feeding any grain except in very small quantities.

The heifer should not be bred under any circumstances until she is at least eighteen months old, and no harm will result from letting her go a few months longer. Too early breeding has a strong tendency to stunt the growth and vigor of the animal and seriously reduce the profitableness to which she may be developed. After the first calf, milk the young cow three times a day, as frequent milking will develop the udder and increase the flow of milk. There is an old saying that "the more you milk, the more you may."

TREATMENT OF SCOURS IN CALVES

In a circular issued by the Wisconsin College of Agriculture, the following method is given for the treatment of scours in calves:

"As soon as symptoms appear, two to four tablespoonfuls of castor oil are mixed with one-half pint of milk and given to the calf. This is followed in four to six hours by one teaspoonful of a mixture of one part salol and two parts sub-nitrate of bismuth. It can also be given with one-half pint of new milk or the powder placed on the tongue and washed down by a small amount of milk.

"The salol and sub-nitrate of bismuth can be secured from any druggist mixed in the proper proportions at the time of purchase and thus have the powder readily available for use at any time. As an additional precaution against contagious scours, it is advised that the navel of the new-born calf be wetted with a 1 to 500 solution of bichloride of mercury (corrosive sublimate)."

Growing into money makers

VENTILATION

Every farmer realizes that moldy, decayed feed is injurious to the health and productivity of his cows. But how many realize the serious effects of forcing the cows to breathe exhausted, impure air? Clean, pure air contains oxygen, which is just as necessary to the cow as hay, grain or water. If the air breathed is contaminated or lacking in oxygen, the results will be just as injurious as feeding moldy, decayed feed.

The lungs are the means by which oxygen is supplied to the blood from the air, and they are also the means by which carbon dioxide is thrown off. The oxygen of the air when breathed into the lungs is absorbed by blood and is necessary to health and life. Carbon dioxide, which the expelled breath carries out of the lungs, is a poison.

Therefore, if the air which has once been breathed, depleted of oxygen and loaded with carbon dioxide, is not removed from the barn the cows will breathe, not oxygen, but instead, poisonous carbon dioxide. The effect of this on the cows will be weakened constitutions, disease, and a reduced flow of milk. To their owner it will mean smaller profits.

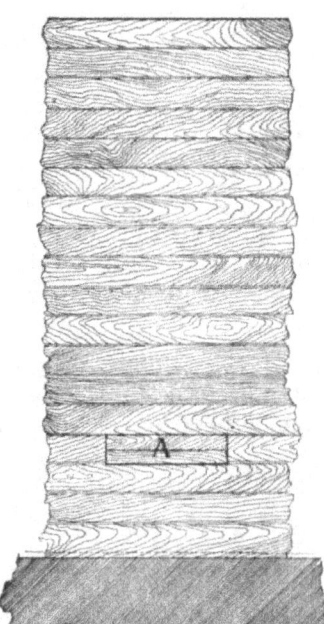

This illustration shows the outside opening of air, intake flue (A) as used in a barn with board walls

The object, then, of ventilation is to bring fresh air into the barn and remove from the barn the air that has been breathed and which contains the poisonous carbon dioxide. The system of ventilation used should be one that accomplishes these results without making the barn cold or causing cold draughts.

There is probably plenty of fresh air in a barn that has broken or open windows, or wide, open cracks, but such a barn will be so cold that most of the feed a cow receives will be consumed in furnishing bodily heat. Warmth is necessary, but it must be warm with pure air.

One of the most satisfactory systems of ventilation is what is known as the King system. In this system two sets of flues

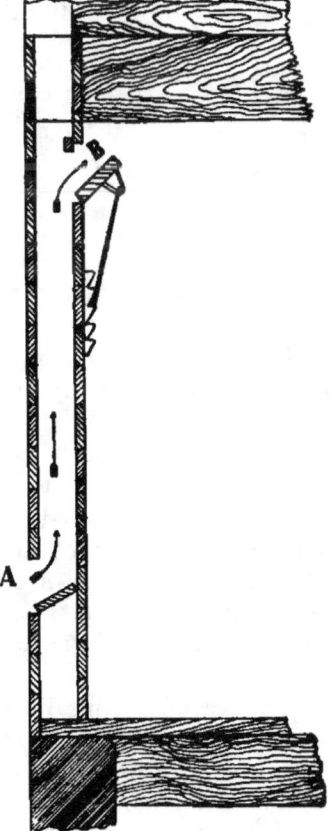

Cross section of a board wall, showing outside opening (A) and inside opening (B) of air intake flues

are used One set admits the fresh air and the other set provides an outlet for the foul air. This system can be installed when the barn is built or it may be installed in barns which were not so equipped when built.

The illustrations show two styles of the intake flues — one for use in barns where the walls are of wood, and the other for use in stone or concrete walled barns.

The flues should be located at least every ten feet along both sides of the barn. The outside openings are located near the ground and the delivery openings inside the barn, near the ceiling. In this way the fresh air that is brought into the barn mingles with the warm air near the ceiling and a large part of the chill is taken out of it before it sinks to a level with the cows.

The openings of these flues through which the air is admitted to the barn should be provided with shutters, so that the amount of air admitted can be regulated. This regulation is very necessary in extremely cold weather, or when a cold wind is blowing directly against the outside opening of the flues.

Concrete wall showing air intake flue outside opening (A)

In barns with wooden walls, these flues can be made by simply utilizing the spaces between the studding. The spaces that are to be used as intake flues, however, should be lined with heavy tar felt paper. In stone or concrete walled barns, the flues are made either of vitrified or of ordinary clay tile. The vitrified tile are much more durable than the ordinary tile, which do not very well withstand the constant action of the air.

The accompanying illustration of a cross section of a barn shows how the foul air flues are installed. These are usually two in number. One is located on each side of the barn midway between the ends of the building. The flues extend from the floor, or near to floor, to the highest point of the

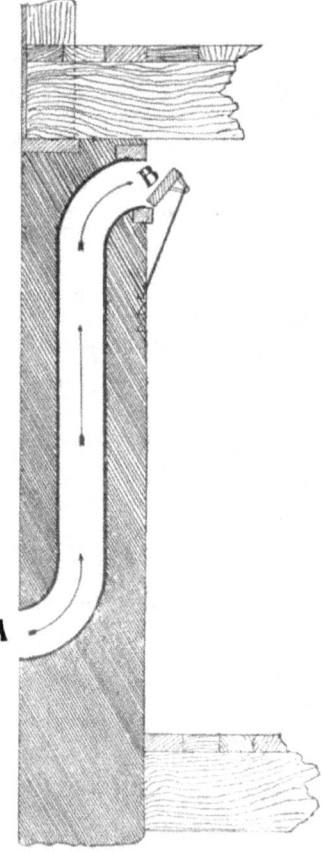

Cross section of concrete wall showing outside (A) and inside (B) openings of air intake flue

building. Bringing the flues close to the floor accomplishes two purposes. First, it removes the carbon dioxide and foul air from the barn. Second, as the cold air is near the floor and the warm near the ceiling, having the flues near the floor removes the cold air instead of the warm. In this way the impure air is disposed of without materially reducing the temperature of the barn.

These flues should be made with as few turns or bends as possible. Galvanized iron or wood may be used in making them; but, if wood is used, the flues should be lined with tar-felt paper.

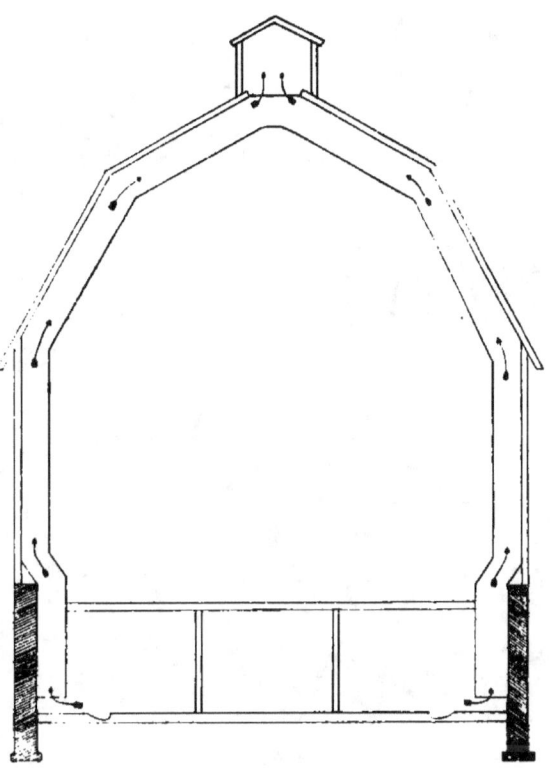

Cross section of a Barn, showing how foul air flues are installed

Interior view of C. S. Sharp's Dairy Barn at Auburn, N. Y., showing a system of ventilation which is a slight modification of that described on the preceding pages. The special construction of the window frames here provides for the intake of fresh air. The principle is the same as the King system

Inside view of C. S. Sharp's Dairy Barn, showing how an abundance of sunlight is admitted

SUNLIGHT THE GREAT DESTROYER OF GERMS

Sunlight is furnished free by nature to preserve the health of all animal life. It is the germ destroyer. It is necessary to admit the sunlight freely to all parts of the stable. For this reason the ridgepole of the barn ought to run north and south to admit the sunlight on the east side of the barn in the forenoon and on the west side in the afternoon.

Big round or square barns with the cows huddled together in masses are bad, so are basement barns in which the sunlight is excluded by the earth on one side or possibly on two.

The barn ought to be long and narrow, not more than two rows of cows being accommodated. These cows may face either toward the center alley or they may face outward.

Of the two methods of arranging the cows, it is difficult to decide which ought to be preferred. Where the cows face in there are no obstructions to the entry of the sunlight which may be allowed to flood the whole floor where the cow stands. If the cows' fasteners and mangers are thrust up toward the windows, they stop the sunlight in great part, and the floors on which the cows stand are kept in perpetual shade.

The floor should be of cement, not troweled smooth, but left somewhat rough so that it may not be slippery when wet. Such a floor is somewhat more expensive at first cost than wooden floors, but its permanent character and the fact that it may be easily cleaned and kept free from odors is enough in itself to decide every dairyman in its favor.

THE VALUE OF SILAGE

It is common knowledge that cows produce greater quantities of milk when fed green, succulent feed than when kept on dry feed. June is the month, in most cases, when the cows yield their largest flow of milk. This is due chiefly to the fact that they have been turned on to fresh, green pastures, where they get an abundance of succulent food. In cold winter months, when the pastures are frozen and covered with snow, silage, properly preserved, furnishes feed that is highly nutritious and keeps the cow's digestion and appetite in the best possible condition. Silage is recognized by successful dairymen as absolutely necessary for economical milk production.

During the summer months, when the cows are on pasture, silage may be dispensed with. The wise dairyman, however, is now supplementing pasturage with silage, and in many cases substituting silage for pasturage entirely. The results are practically as good, and the cost much less. This is an important consideration, as very few farmers can afford, because of the high value of farm land, to set aside a large enough acreage of pasture to properly feed their cows. It is not necessary that they should, because on a comparatively small amount of land they can raise enough silage corn to feed their herds. Besides, pastures are uncertain. A few weeks of dry weather will make them worthless.

The superiority of corn silage over dried corn fodder lies in the fact that the silage is juicy and appetizing; cattle relish it when they would reject ordinary dried fodder. The Vermont Experiment Station made a careful test of the relative values of corn silage and dried fodder. The results were:

24,858 pounds of green fodder corn, when dried and fed with a uniform allowance of hay and grain, produced 7,688 pounds of milk.

24,858 pounds of green fodder corn, converted into silage and fed with the same ration of hay and grain as was fed with the dried fodder, produced 8,525 pounds of milk.

The following is quoted from Prof. W. A. Henry's Feeds and Feeding: "Indian corn is pre-eminently suited for silage. The solid, succulent stems, when cut into short lengths, pack closely and form a solid

A good crop for the silo

mass, which not only keeps well, but furnishes a product that is greatly relished by stock — especially cattle. It is reasonable to estimate that there are over 100,000 silos now in use in America. Probably 95 per cent of all the forage stored in them is from the corn plant, and 95 per cent of the silage made is fed to dairy cows."

WHEN AND HOW TO FILL THE SILO

The quality of silage is determined to a great extent by the condition of the corn when cut, and the care used in filling the silo. The question of the proper time to cut corn and store in the silo has been much discussed and studied. Experience and careful study of results show that the best silage is made from corn that is cut and put into the silo at the time the kernel dents and begins to harden. At this stage the corn has practically attained maturity and its full nutritive value is developed. This, of course, will depend to some degree on the kind of corn which is grown. Some varieties of corn dent easier than others, and care should be taken not to let the corn become too dry. Dry corn fodder does not pack and exclude the air as well as that which contains a considerable amount of moisture. If the use of over-dry fodder cannot be avoided, its disadvantages can be overcome, in a measure, by adding water as the corn is put into the silo. The amount of water to be added is a matter of judgment, and the person applying it should have had some experience with silage. Too much water will cause the silage to develop an excessive amount of acidity. On the other hand, if enough water is not used, the silage will not settle properly and exclude the air; this will cause moldy silage.

A great many farmers make the mistake of cutting their corn and putting it into the silo in a very immature state. This is often due to the fear that an early frost will injure the corn from a silage standpoint. Of course, being frosted does not improve the quality of the silage, but the damage due to a slight frost is not great. It is much better to take a chance of the corn being slightly frosted than to put it into the silo in a green, immature condition. Many times the expected frost will not come, but the quality of the silage will always be reduced by using corn

Filling the silo

that is not sufficiently developed. Too green corn, if put into the silo, will make silage that has an excessively high amount of acid and a reduced feeding value. The cattle will eat it fairly well, but will not relish it as much, nor receive as much nourishment from it as they would from silage made from more mature corn.

When the silage is elevated into the silo by the blower or elevator, it is not evenly distributed, and consequently it is a good plan to have a man in the silo to fork the silage about. This will insure an even distribution of the light and heavy parts of the silage. The man in the silo can also devote part of his time to tramping the silage, especially around the edges. Tramping and packing the silage will add greatly to its keeping qualities by excluding the air.

The cost of filling a silo has been estimated to average about 56 cents per ton, the range of cost being from 40 to 76 cents. The difference is due to the distance the corn must be hauled, the experience and skill of the men doing the work, and the size and power of the machines used.

SUGGESTIONS FOR BUILDING SILOS

There are many types of silos, and most all of them are good. The kind of silo to be built must be determined by local conditions and personal preference. A number of the most common types are shown on this and the following pages. As a rule, it will pay in the long run to build the most substantial silo.

No matter what type of silo is selected, it should be well built. The silo must be perfectly airtight, and substantial enough so that the pressure of the silage will not cause it to bulge. It must be fairly deep, so that the weight of the silage will cause pressure enough to exclude the air.

A wood silo of the stave type

One of the first things to be considered when about to build a silo is the size necessary. The diameter of the silo should be such that, when feeding, three or four inches of silage will be removed from the top every day. This is important, as the silage when exposed to the air molds and becomes unfit for feeding, but if three or four inches are removed evenly from the top each day, the silage will not be exposed to the air long enough to become damaged. If more silage is needed than can be stored in a silo twenty feet in diameter by fifty feet high, a second silo is preferable to a silo larger than the above dimensions. If the silo is more than twenty feet in width, it is probable that enough will not be removed at each feeding to prevent molding. It is impracticable to elevate the silage more than fifty feet, as this is about as high as it can be conveniently elevated with ordinary farm power. Besides, the weight of the silage exerts great pressure against the sides of the silo, and if the silo is built extremely high there will be danger of bulging at the bottom, where the pressure is greatest.

The silo should be built right up against the barn in which the silage is to be fed, preferably at the end of the building, as this will be found most convenient for feeding. There is no reason why the silo should be set off at a distance from the barn, and if this is done the work of bringing the silage to the barn will be considerable. The construction of the silo should not be undertaken by inexperienced persons. Silo building

Two large cement silos

has become quite common in many localities, and it will not be hard to find help who have had some experience in this work.

The foundation should be solid and well-made, as is true regarding the foundation of any permanent building. As the weight placed upon the foundation is that of the upper part of the silo, the necessary thickness of the walls will depend upon the material used in constructing the part of the silo that is above ground. A concrete silo will naturally require a thicker foundation than a light stave silo. The foundation should extend below the frost line, and the ground in which it sets should be well drained. If the foundation must be built in soil which contains a considerable amount of moisture, the foundation and floor should rest on a bed of gravel or cinders, and drain tile should be provided to carry off the water.

The inside walls should be smooth and perfectly perpendicular, so that the silage can settle evenly without sticking on the walls. The walls must be air-tight and water-tight. Air will cause the silage to rot, and loss of moisture will cause it to become dry and moldy. Not only should the walls be water-tight, but they should be constructed of material which will not absorb the water from the silage, as this will dry the silage and cause molding just the same as if the water leaked out.

The capacity of round silos, which are the only kind that should be built, is not readily computed, but the table on the following page has been prepared so that an approximate capacity can be seen at a glance. This table includes silos 10 to 26 feet in diameter and from 20 to 32 feet high.

Two wood silos, one lined with vitrified brick and the other with lath and plaster

Table Giving the Approximate Capacity in Tons of Cylindrical Silos for Well Matured Corn Silage

Depth of Silo—feet	Inside Diameter of Silo—Feet												
	10	12	14	15	16	18	20	21	22	23	24	25	26
20	26	38	51	59	67	85	105	115	127	138	151	163	177
21	28	40	55	63	72	91	112	123	135	148	161	175	189
22	30	43	59	67	77	97	120	132	145	158	172	187	202
23	32	46	62	72	82	103	128	141	154	169	184	199	216
24	34	49	66	76	87	110	135	149	164	179	195	212	229
25	36	52	70	81	90	116	143	158	173	190	206	224	242
26	38	55	74	85	97	123	152	168	184	201	219	237	257
27	40	58	78	90	103	130	160	177	194	212	231	251	271
28	42	61	83	95	108	137	169	186	204	223	243	264	285
29	45	64	88	100	114	144	178	196	215	235	256	278	300
30	47	68	93	105	119	151	187	206	226	247	269	292	315
31	49	70	96	110	125	158	195	215	236	258	282	305	330
32	51	73	101	115	131	166	205	226	249	271	295	320	346

Referring to the table below, it will be an easy matter to determine the size of the silo needed. This table is based on the assumption that 40 pounds of silage will be fed per head for a period of 180 days:

Number of Cows	Pounds Required Daily	Silage Consumed Yearly—Tons	Size of Silo Needed			Acreage Required at 15 Tons per Acre
			Diameter Feet	Height Feet	Capacity Tons	
6	240	22	9 / 10	20 / 16	22 / 22	1.5 / 1.5
9	360	33	10 / 11	24 / 22	34 / 34	2.4 / 2.4
13	520	42	10 / 10	28 / 30	42 / 47	2.8 / 3.0
15	600	54	12 / 14	26 / 21	55 / 55	3.7 / 3.7
20	800	72	12 / 14	32 / 26	74 / 74	5.0 / 5.0
25	1000	90	12 / 14	38 / 30	94 / 91	6.4 / 6.1
30	1200	108	14 / 15	34 / 31	109 / 110	7.3 / 7.4
35	1400	126	16 / 14	31 / 38	125 / 128	8.4 / 8.6
40	1600	144	18 / 16	29 / 34	144 / 143	9.4 / 9.3
45	1800	162	18 / 16	32 / 38	166 / 167	11.0 / 11.1
50	2000	180	18 / 16	34 / 40	181 / 180	12.1 / 12.0

THE BABCOCK TEST

The Babcock test has been one of the chief factors in demonstrating the fact that too large a percentage of dairy cows are kept at an actual loss to their owners. With milk scales and the Babcok test, a farmer can learn just what each cow in his herd is producing. In this way he can easily locate and cull out those cows which do not return a good profit or those which are not paying for their feed. Weighing the milk is not sufficient, as the milk from different cows varies greatly in percentage of butter fat, and it is butter fat that determines the market value of milk. Hence, the Babcock test is of immeasurable value to the man who keeps milk cows. It gives him a simple, reliable means of ascertaining which cows in his herd are producing enough butter fat to make it worth while to keep them. Farmers who do not use this means of finding out what their cows are doing usually make the excuse that it is too much trouble. As a matter of fact, the work of keeping these records is not nearly so great as it may seem. Even if it were a great deal more trouble than it is, it would be better to put in time finding the unprofitable cows and getting rid of them than to go on feeding and milking cows that do not produce enough to pay for their feed and care.

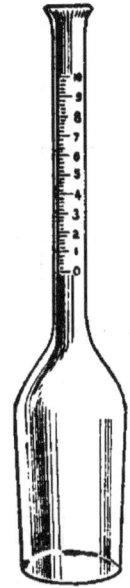

Milk Test Bottle

HOW TO KEEP A RECORD OF EACH COW'S PRODUCTION

In keeping a record of the milk and butter-fat production of a herd there is needed: a spring balance scale, pint glass jars, test bottles, pipette, acid measure, a bottle of sulphuric acid, preservative tablets, a

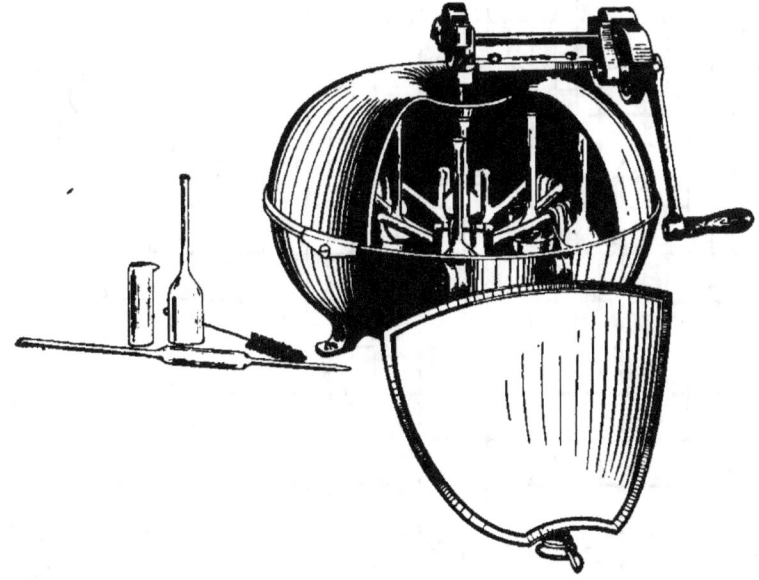

A Babcock Milk Testing Outfit

FARM MILK RECORD

Month.................... Year.................... Owner's Name....................

COW'S NUMBER	1		2		3		4		5		6		7		8		9		10		11		12		13		14		15		16	
DATE	AM	P.M	AM	P.M	AM	P.M	AM	P.M	AM	P.M	AM	P.M	AM	P.M	AM	P.M	AM	P.M	AM	P.M	AM	P.M	AM	P.M	AM	P.M	AM	P.M	AM	P.M	AM	P.M
1																																	
2																																	
3																																	
4																																	
5																																	
6																																	
7																																	
8																																	
9																																	
10																																	
11																																	
12																																	
13																																	
14																																	
15																																	
16																																	
17																																	
18																																	
19																																	
20																																	
21																																	
22																																	
23																																	
24																																	
25																																	
26																																	
27																																	
28																																	
29																																	
30																																	
31																																	
Total AM & PM																																	
Total																																	
Test																																	
Lbs Butter Fat																																	
Value																																	

centrifugal machine, and a sheet for recording the weight and test of each cow's milk. The record sheet, ruled as shown on the opposite page, should be placed with the scales in a convenient position in the barn and the milk of each cow weighed at each milking, and the weight recorded on the sheet.

The testing of the milk for butter-fat can be done daily, weekly, or monthly. The practice of making the test once a month meets most requirements. The monthly test does not involve so much work as more frequent tests, and is a very good indication of the per cent of butter-fat the cow is producing.

In making this test, samples should be taken from each milking for a period of three days and placed in pint glass jars. To prevent the samples from souring, a small corrosive sublimated tablet should be put into the jar. A small dipper (about the size of a shotgun shell) with a long handle proves most satisfactory for taking the sample. Before taking the sample, the milk in the pail should be well stirred with the dipper.

The testing should be done as soon as possible after the samples from six milkings have been taken. The operations of this test are as follows: First. The samples should be stirred by pouring into and out of an extra jar several times. In making the Babcock test, 17.6 C. C. of milk is used and is measured by means of a pipette, which is marked to show when this amount is in it. When using the pipette, place the small point in the milk and with the other end in the mouth suck the air out of the pipette until the milk rises above the 17.6 C. C. mark. Then quickly place the tip of the forefinger over the end of the pipette which has been in the mouth. This will prevent the milk from running out of the pipette. By slightly changing the pressure of the finger on the end of the pipette, the milk can be allowed to run down slowly until the 17.6 C. C. mark is reached. Then press the finger firmly on the end of the pipette to prevent any more of the milk from running out.

Pipette

Second. When exactly 17.6 C. C. of milk are contained in the pipette, place the small end of the pipette in the top of the test bottle and gradually reduce the pressure of the finger on the other end. The pipette should not be put straight down into the test bottle; instead, the bottle and pipette should be at a slight angle so that the milk will flow down one side of the neck of the bottle and at the

Showing proper way to hold Pipette and Test Bottle when filling Test Bottle

same time leave a space on the other side for the escape of the air which the milk displaces. Don't allow the milk to run out of the pipette too fast or it will choke the neck of the bottle and overflow. This would require washing the bottle and measuring a new sample of milk with the pipette.

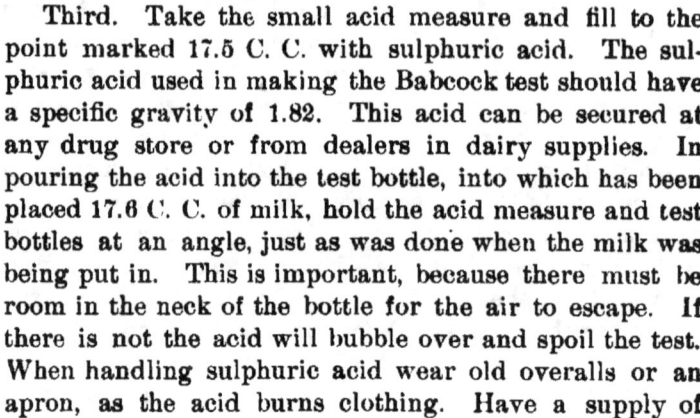
Acid Measure

Third. Take the small acid measure and fill to the point marked 17.5 C. C. with sulphuric acid. The sulphuric acid used in making the Babcock test should have a specific gravity of 1.82. This acid can be secured at any drug store or from dealers in dairy supplies. In pouring the acid into the test bottle, into which has been placed 17.6 C. C. of milk, hold the acid measure and test bottles at an angle, just as was done when the milk was being put in. This is important, because there must be room in the neck of the bottle for the air to escape. If there is not the acid will bubble over and spoil the test. When handling sulphuric acid wear old overalls or an apron, as the acid burns clothing. Have a supply of water convenient to wash off any acid that may spill on the hands or clothing.

Fourth. As soon as the acid has been poured into the test bottle with the milk it will be noted that the milk and acid lay in two distinct layers—the acid in the bottom of the bottle and the milk on the top of it. The immediate mixing of these two layers is important. Do this by taking the bottle by the neck and swinging it in a circle until acid and milk are completely mixed. This mixture has a uniform brown color and becomes very hot. On the rough spot on the side of the test bottle write with an ordinary lead pencil the cow's number whose milk is being tested, or write some number that will serve as a means of identifying the bottle.

Whirl the test bottle in a circle to mix the acid and milk

Fifth. After the milk and acid are thoroughly mixed, place the test bottle, together with other test bottles which have been filled in a similar manner with the milk of other cows, into the centrifugal or whirling machine. After making sure that the bottles are so placed in the machine that they balance, turn the crank four or five minutes at the speed indicated in the directions supplied with the machine.

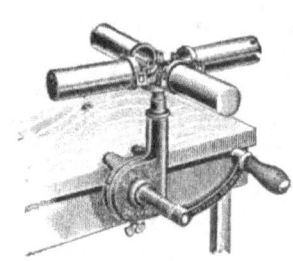
A small four-bottle Babcock tester

Sixth. After whirling the bottles in the machine four or five minutes, stop turning and allow them to gradually come to a stop. Then take the pipette and add to each bottle, without taking it out of the machine, a small amount of hot soft water. The water put into the bottles should come to the bottom of the neck or a little above it. Then start the machine again, and whirl the bottles for at least two minutes. Next add enough more hot water to bring the fat which has gathered at the bottom of the neck to a point between the top and bottom figures of the scale on the bottle. Whirl for one minute more.

Seventh. Remove the bottle from the centrifugal machine and proceed to read the per cent of fat in the neck of each bottle. It is important that the reading be made while the fat is hot, therefore set the bottles in a dish of water at the temperature of 130 or 140 degrees. The scale on the Babcock test bottle is graduated from 0 to 10 per cent. The scale on the neck between 0 and 10 is divided into 10 spaces, each representing 1 per cent of fat in 100 pounds of milk. Each of these spaces is subdivided into 5 equal parts, each representing .2 of 1 per cent. If the fat found in the neck after the whirling has been completed extends from 0 to 4 it means that the milk tested contains 4 pounds of butter fat for every 100 pounds of milk, or, in other words, the milk tests 4 per cent butter fat. It is not very often that the bottom of the fat column will be formed exactly at the point marked 0, and in most cases it will be somewhat above this point. Hence, the work of reading can be greatly facilitated if a pair of dividers is used. In using the dividers adjust the points to the top and bottom of the fat column and then, without changing the distance between the points, place one point on 0 and read on the scale the percentage of fat which is indicated by the position of the other point.

The reading of the test should be made from A to B, that is between the extreme top and bottom of the fat column

First position of the dividers when used for reading the test

Second position of the dividers when used for reading the test

Dolly Bloom, Guernsey Cow. Has a record of 17,297.5 pounds milk in one year, testing 4.87 per cent, yielding 836.2 pounds butter fat

TESTING SKIM MILK

When testing skim milk, a double-neck skim milk test bottle should be used, as it gives a better reading. About 20 C. C. of acid should be used, as in skim milk there is a larger amount of solids not fat than in whole milk. These must be destroyed before the fat can be freed. Otherwise, the operations are the same as for testing whole milk.

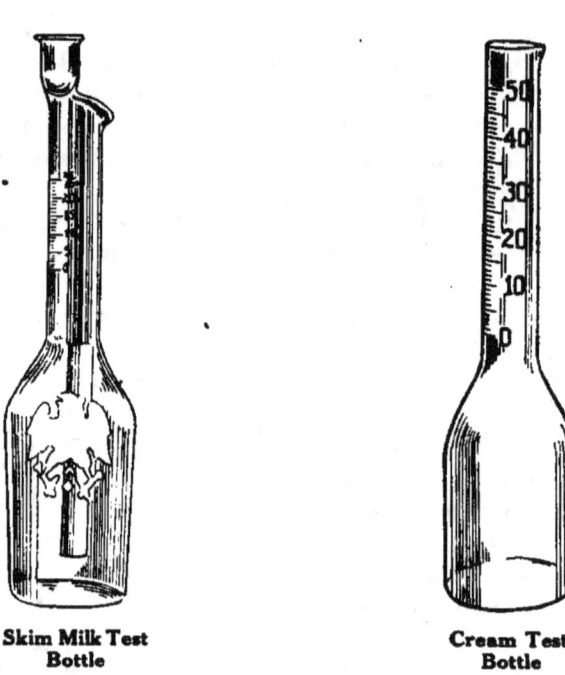

Skim Milk Test Bottle

Cream Test Bottle

TESTING CREAM

The operation of testing cream with the Babcock test is the same as for testing milk, with two exceptions: First, a special cream test bottle should be used. This cream test bottle has a larger neck than the milk test bottle. This is because the amount of fat in cream is much greater than in milk. Second, the 18 grams of cream used in making the test cannot be measured with the pipette, but instead must be weighed. This is due to the fact that the weight of cream varies according to its richness. Furthermore, cream is thick and a considerable part would stick to the inside of the pipette. There is also another objection to measuring the cream with a pipette, and that is that cream, especially fresh separator cream, often contains bubbles. Therefore, to get an accurate test the cream must be weighed. There are scales made especially for this purpose, and these can be secured from dealers in dairy supplies. The remainder of the operation for testing cream is the same as for testing milk. The testing of cream is much more difficult, however, than the testing of milk and considerably more experience is necessary to make a good cream test.

THE COMPOSITION OF MILK

The composition of milk varies greatly, depending upon the breed and individuality of the cow, the season of the year, lactation period, milking, and environment. The average composition, however, which has been determined by 200,000 analyses reported by a well-known dairy authority is as follows:

Water	87.17
Fat	3.69
Milk Sugar	4.88
Casein	3.02
Protein, Albumen	.58
Ash	.71

COMPOSITION OF SKIM MILK

When cream is taken from the milk by a separator or by hand, practically all of the fat is taken out. The skim milk which remains is frequently referred to as "serum," and it contains everything but the fat, as follows:

Water	90.68
Fat	.02
Milk Sugar	5.00
Casein and Albumen	3.50
Ash	.80

COMPOSITION OF BUTTER

Butter is composed of fat, water, proteids, milk sugar, ash, and salt in the following average proportions, according to a well-known dairy authority:

	From Fresh Cream	From Ripened Cream
Fat	83.75	82.97
Water	13.03	13.78
Proteids (Curd)	.64	.84
Milk Sugar	.35	.39
Ash	.14	.16
Salt	2.09	1.86

The quality of butter is more affected by the quality of cream or milk from which it is made and the methods employed in manufacture than by the composition.

The English, German and United States governments endeavor to protect the consumer of butter by recommending 16 per cent of water as a maximum limit. Butter is frequently found which contains more than 16 per cent of water, but this is in violation of the law. The amount of fat in the butter varies with the water—the more water, the less fat there will be. Butter which contains more than 18 per cent of water will appear dead and dull. It will also be leaky.

STANDARD FOR JUDGING BUTTER

In judging butter, the different characteristics are given different values according to their relative importance. Below is given a standard used commercially and based upon 100 as perfect:

	Perfect
Flavor	45
Body	25
Color	15
Salt	10
Style	5
	100

FLAVOR. — As shown in the score above, flavor is the most important characteristic. Good butter should possess a clean, mild, rich, creamy flavor, and should have a delicate, mild, pleasant aroma. FLAT flavor is noticeable in butter made from unripened cream. RANCID flavor is applied to butter which has a strong flavor, and develops in butter which has been standing a long time. CHEESY flavor is common to butter which has little or no salt. WEEDY flavors are due to the condition of the milk before churned and are caused by the cows pasturing where weeds are growing, such as wild onions, garlic, etc. ACID flavor is due to improper ripening of the cream.

BODY. — Next in importance to flavor is body. Butter that is greasy, tallowy, spongy, or sticky is undesirable. The body must be firm and uniform.

COLOR. — The color should be bright and even, not streaky or mottled. A light straw color is the color most desired.

SALT. — The amount of salt depends upon what the market wants. The principal thing is to have the salt thoroughly dissolved and evenly distributed. Medium salting is most desired.

STYLE. — By style is meant the appearance of the butter and package. It should be clean and neat.

Colantha 4th's Johanna. Holstein-Friesian Cow which produced in one year 27,432.5 pounds milk testing 3.64 per cent, yielding 998.25 pounds butter fat.

This Holstein-Friesian Cow, Banostine Bell De Kol, holds the world's record for butter production, having produced in one year 27,404 pounds of milk containing 1058 pounds of butter fat

BUTTER ON THE FARM

It is possible for the farmer to make the highest possible grade of butter on the farm, owing to the fact that he has the entire control of the milk from the time it is drawn until it is turned out a finished product ready for the market. Especially is this true where the farmer has a small separator.

With proper ripening before churning, and careful observation of necessary conditions for the production of the best butter, the farmer should be able to economize in the making and insure a distinct saving by feeding the by-products to the pigs and calves.

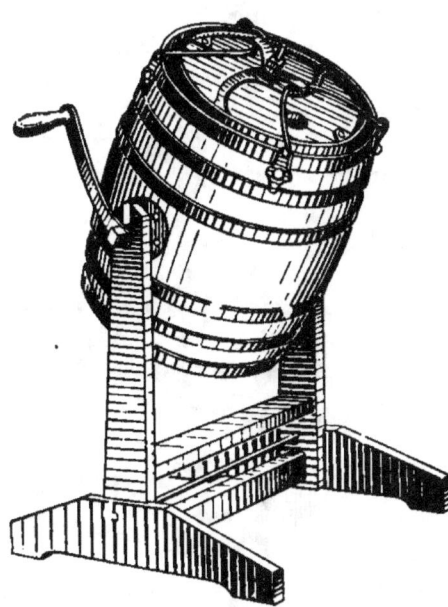

This style of churn proves very satisfactory on the average farm

If the farmer can furnish an even grade and a regular supply of butter the year round, he can with a little effort readily find an excellent market. People living in towns and cities generally prefer to buy butter direct from the farmer, if possible, and are willing to give the farmer his price.

Besides the income derived from butter sales, the by-products fed to pigs and calves are steadily increasing the value of young beef and pork. The buttermilk, if fed directly after churning, is always productive of good results, as the chances of fermentation or contamination are fewer than in the creamery and consequently it gives better results as a feeding ration.

CLEANLINESS

Volumes and volumes could be written on the subject of cleanliness and its relation to the dairy. There are thousands of arguments in its favor and the statistics covering the point cannot well be ignored by the farmer or the dairyman.

Clean cows, clean udders, clean hands, clean pails, sterilized utensils and separators, clean and thoroughly ventilated, sweet-smelling dairies — these are some of the conditions under which milk, cream and butter can be best preserved and utilized for home use and for the market.

Do not stir up unnecessary dust before milking. Each minute particle of dust settling on the milk means that much taint and consequent germination of bacteria. All strainers should be kept scrupulously clean. Sanitary wire gauze strainers are greatly to be preferred to the common cloth strainers so much in vogue. All foreign odors should be abolished from the premises, as milk, cream and butter have a natural tendency to absorb bad smells.

The stable should be provided with brushes readily attached to the milking stools or accompanying them. The milker should be encouraged to use these brushes before milking, and if such milkers are naturally cleanly, they should also be encouraged to dampen the udders before beginning to milk. If the milkers are not naturally orderly, systematic and cleanly, discharge them and either get clean milkers or quit the business. It is impossible to make a filthy man clean by any set of rules or by any amount of possible supervision. "Though thou shouldst bray a fool in a mortar among wheat with a pestle, yet will not his foolishness depart from him."

The milk is received in pails washed in this way: They are first rinsed in tepid water, then washed in water too hot for the hands and containing some cleansing powder or sal soda, the washing being done with brushes rather than cloths. They are then rinsed with boiling water and steamed, if possible; otherwise they are taken from the rinsing water, the loose drops shaken off and allowed to dry without wiping. The milk is then strained through wire strainers or through two or three thicknesses of cheesecloth, which pieces are washed and scalded or boiled between successive hours of milking.

After straining, the milk is either aerated, cooled and sent to the factory, or it is run through the separator at home.

The cream separator is one of the best milk clarifiers. It removes the finest particles of dirt from the milk which could not be removed by a cloth or wire strainer. Even if the whole milk is to be sold at retail, it should be run through the separator for clarifying purposes.

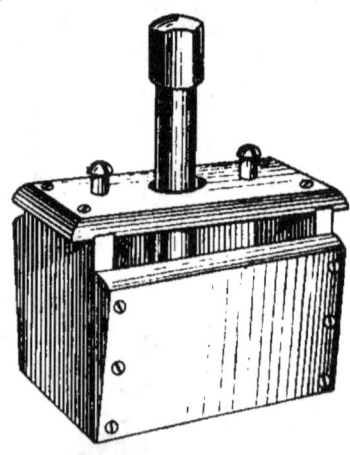

Butter sells better when put up in attractive prints. A butter mold similar to the one illustrated here can be secured at a small cost

Missouri Chief Josephine, a Holstein cow with a year's record of 26,861 pounds of milk containing 740.5 pounds of butter fat

CREAM RIPENING AND STARTERS

Cream ripening is generally understood to mean the treatment and process which the cream undergoes in the ripening vat before it is put into the churn; a process which secures to the butter that fine flavor and scent which is so highly desirable and prized in all good butter. This ripening is caused by the bacteria contained in the cream which produce certain acids through decomposition. It is generally believed by progressive dairy scientists that lactic acid-producing bacteria are most desirable for ripening purposes. There are many species, however, all of which vary in results when applied to the ripening process. Over one hundred species have been analyzed and studied. Wide experiments show that the best temperature for ripening is between sixty and seventy degrees.

The "starter," used in a dairy sense, is the name given to the medium which contains the greatest number of desirable and active bacteria for producing the best flavors in butter.

Some starters come from the laboratory in a liquid or powder form. The most common, however, are the so-called natural starters so much in use, such as buttermilk, sour milk, and sour cream. These latter are not perhaps the best, but serve the purpose and give better results than if no starter is used at all.

The flavor of butter is largely controlled by the kind of bacteria that predominate in the cream. Certain groups of bacteria are known as flavor-producing or lactic acid bacteria. Other groups are known as putrefactive bacteria or those that cause the ordinary decay. During the winter months when cows are milked in the stables the latter kind seem to predominate in the milk and give the butter an undesirable flavor. Germs get into the milk from an external source, coming from manure and from the atmosphere, hence very poorly ventilated barns might be called incubators for undesirable bacteria.

To overcome the effects of the latter, a starter should be used. Most creameries use commercial starters that are prepared in laboratories. A good, natural starter that will answer the purpose equally well can be prepared on the farm. Put pint glass fruit jars into cold water and let the water gradually come to the boiling point. This will sterilize the jar or destroy all germs in the jar. After the jars have cooled, close them until time for using. At milking time carefully brush and dampen the udder, then after a few streams of milk have been drawn, milk directly into your sterilized glass jar, filling it half full. Cover the same and let it stand, keeping the temperature as near 65 or 70 degrees as possible until the milk coagulates or thickens. If the curd in the glass jar is free from pin holes or little openings in the sides and possesses a pleasant sour taste, you can be sure that you have a good starter. This we call the "mother starter." To propagate it or prepare it for the cream, put into a clean tin pail a few quarts of skimmed milk from the separator and heat the same to about 170 or 175 degrees. The heating can best be accomplished by putting the pail of skimmed milk into a larger vessel containing hot water, thus preventing any danger of scorching the milk. Cool your milk now to about 70 degrees and then add to it 2 or 3 per cent of the mother starter from the glass jar above mentioned. Keep the mixture or starter at a temperature of 65 or 70 degrees until it begins to coagulate or sour. Add to your cream from six to ten pounds of starter to every 100 pounds of cream. Stir the starter thoroughly through the cream. You are thus adding to your cream an enormous quantity of the right kind of bacteria which will enable you to control the flavor of the butter.

As soon as the starter thickens it should be kept at as low a temperature as possible to prevent further souring. A new starter can be made from time to time by preparing fresh skim-milk as above described and adding to it a little of the previous starter.

Dairymaid of Pinehurst, a Guernsey Cow with a year's record, when three years old, of 14,571.4 pounds of milk and 852.82 pounds of butter fat

CHURNING

The primitive method of churning was to shake the milk without separation in bags made from animal skins, preferably goat skins. More has been done in the way of improving upon the system in the last fifty years than during the previous 5,000 years. In Europe not so long ago, churning in many places consisted of shaking cream in glass bottles, jars, or other convenient receptacles. This necessarily was very fatiguing, but a marked improvement was made with the introduction of the dash churn, in which the cream was agitated by the up and down movement of the long handle, to which was attached, at the lower end, a round perforated plate of wood or some other material. This churn was a direct forerunner of the rotary churn now so widely used in Europe.

The most popular churn in the butter factories is the so-called "combined churn," a strictly modern and up-to-date device which churns, washes, salts, and works the butter without necessitating its removal from the churn. The movement of the churn serves to keep flies at a distance; the handling of the butter during working and salting is done away with, and the temperature of the butter can be regulated at will. This style of churn is rapidly taking the place of other devices in Europe, and its many excellent features are generally recognized in butter-producing districts.

In churning, a medium high temperature should be observed for securing the best results. If the temperature is too high, a soft, lumpy butter is the result, which appears greasy to the touch and is very susceptible to the incorporation of buttermilk in larger quantities than desired.

Too low a temperature is also to be carefully avoided, as churning then becomes extremely difficult. The cream will also adhere to the inside of the churn and the butter will become too hard for taking up the salt readily. The difficulty which is sometimes experienced in getting the butter to break, although not only the temperature but all other

This style of butter worker is used on many farms

conditions are favorable, can easily be remedied by adding a little salt to the cream.

With a moderately high temperature and the churn two-thirds full, quick churning is insured and the highest possible degree of agitation then obtained. The agitation is evenly distributed, while on the contrary with a small amount of cream, much of it will stick to the revolving surface or churn walls, thus hindering rapid churning.

It is not advisable to stop churning before the butter flakes or kernels have attained a size which will prevent them from being strained into the buttermilk. One prominent authority says that the butter granules should be of the size of corn kernels.

The butter should be washed thoroughly in pure water of about the temperature of the cream while being churned. The churn should be cleaned, washed and rinsed, first in lukewarm and then in scalding hot water, and finally disinfected with slacked lime in a liquid condition, which is considered one of the best disinfectants that can be employed in the dairy. Salt, on account of its corroding effects on iron, should never be used. Never stand the churn with the cover hole up when drying, as there will always be more or less dust and impurities collected in this way.

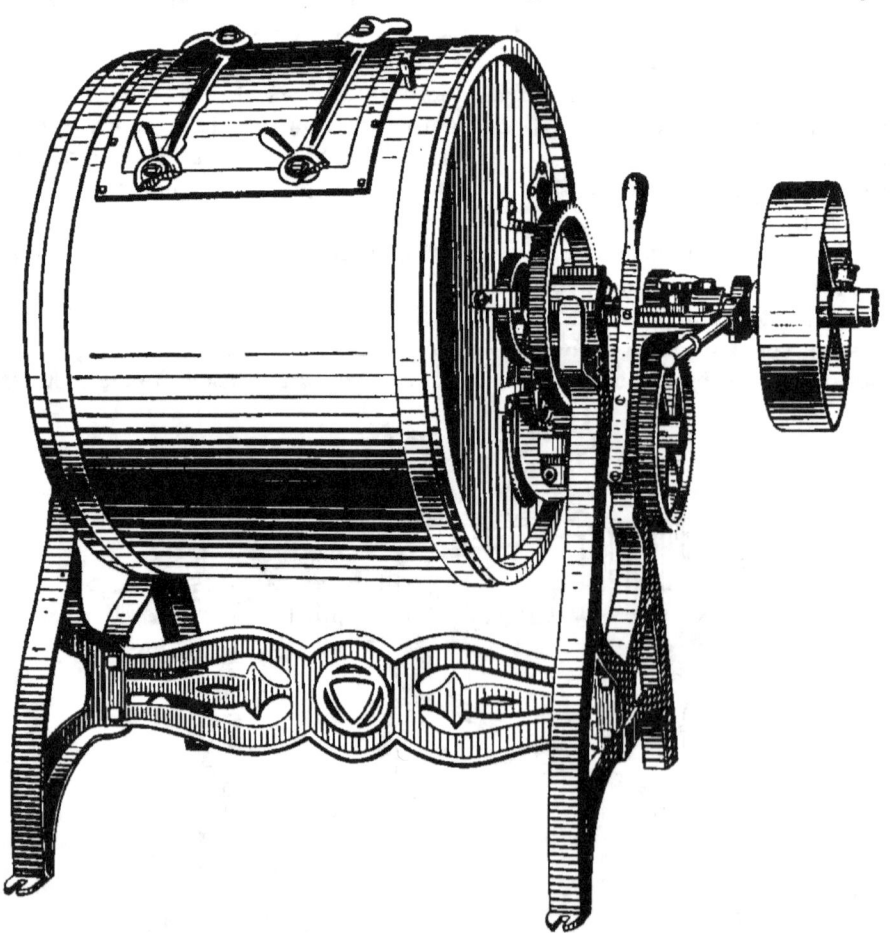

This illustration shows a combined churn and butter worker. This machine has many advantages over separate churn and butter worker for the farmer who churns from 50 to 100 pounds of cream at a time

A practical Dairy Farm Power House

SUGGESTIONS FOR BUILDING A FARM POWER HOUSE

That a gasoline engine is a necessary part of the dairy farm equipment is now conceded. Many of the most progressive farmers who own gasoline engines have found that by installing an engine in a farm power house much larger profits are made; in fact the engine can be made to pay for itself.

In a power house the engine is the most important machine, and should be installed in a room which can be shut off from the remainder of the power house. The engine should be so placed that all sides of it are accessible — at least three feet from any wall — and should be in a well-lighted, well-ventilated room that can be heated in the winter time. If the engine is crowded into a corner, it is impossible to get at all the parts, either for cleaning or repairing, so that the outfit proves inconvenient to care for.

The gasoline tank should be located outside of the engine room about thirty feet distant. If this tank is properly installed in the ground, it will be insurance against fire risk, and will be kept cool in the summer, so that the gasoline will not evaporate.

In the engine room might be located the grindstone, emery wheel, drill and work bench so that this room becomes a work-shop. If a dynamo is used to furnish light for the house, it can be located in this room.

The remaining space in the power house should be divided into two rooms. In one room locate the grinder, sheller, cutter, and fanning mill. The grinder and sheller might be located close to a window so that it is possible to unload from the wagon through the window into the machine. The room containing these machines should be tightly partitioned off from the rest of the building because these machines when in operation create dust, which would interfere with the successful operation of other machines.

In the third room can be located the churn, cream separator, pump and drain, dairy table and a large washing trough. This room then becomes the dairy.

All these machines in the power house should be driven by belt from line shafting which is set overhead or hung from the ceiling.

When an engine of four-horse power or less is installed, 1½-inch iron shafting will be found adequate. If a larger engine is installed, 2-inch iron shafting should be used. Shafting, pulleys, hangers, and belts can be purchased in all towns and small cities from the hardware merchant, while in large cities these supplies are furnished by iron and steel supply houses.

All high speed machines should be firmly bolted to the floor and operated by their own belts, while other machines can be moved so that one or two belts will answer for driving all of them; that is, if they are not to be driven simultaneously. Locate the machines as near the line shafting as possible in order to avoid long belts. The center line of the engine must be exactly at right angles to this shafting, and to obtain the best results the pulley of the engine must be in line with the pulley which is to receive the power. The distance from the line shafting to the crank shaft of the engine should be at least from six to eight times the diameter of the larger pulley.

The line shafting can be extended through the end of the building so that a blacksmith bellows, or circular cut-off and rip saw, band saw, or any other wood-working machinery can be driven on the outside from the same power. The accompanying illustration shows an arrangement which has proved very practical. The cost of such an installation compared to the benefit derived is very slight.

An I H C Hopper-Cooled Gasoline Engine mounted on skids

THE VALUE OF THE CREAM SEPARATOR

Butter fat, the substance that gives milk its market value, exists in milk in the form of minute globules and is the lightest part of the milk. When milk is allowed to stand several hours in pans, cans or crocks, the action of the force of gravity causes the heavy milk serum to sink to the bottom and the butter fat rises to the top as oil rises to the surface of water.

Methods of separation which depend upon the unaided action of gravity are unsatisfactory because the action of gravity is not strong enough to bring about a complete and rapid separation.

In a bulletin published by the Illinois Experiment Station, Carl E. Lee, assistant chief dairy manufacturer, says:

"The old methods of putting milk in shallow pans in a cool place, or in deep cans in a tank of cold water, are still in use, but not all the cream is recovered by these methods. However, they are more satisfactory than the so-called water separator, which is nothing but a fraud—a piece of apparatus deceiving to the user. * * * By this hydraulic, or water separator, from one-fifth to one-fourth of the butter fat is lost. Skim milk of low feeding value is obtained and the cream is thin and often contaminated with all the impurities of the water. * * * * The most satisfactory method of obtaining cream from milk on the farm is by the use of a standard make of hand separator. The cost of such a machine may seem high, but when the amount of butter fat is compared with the butter fat obtained from the same milk by other methods, one can easily figure how long it will take to save the cost of the machine."

Butter Lost in Skim Milk from One Cow in One Year

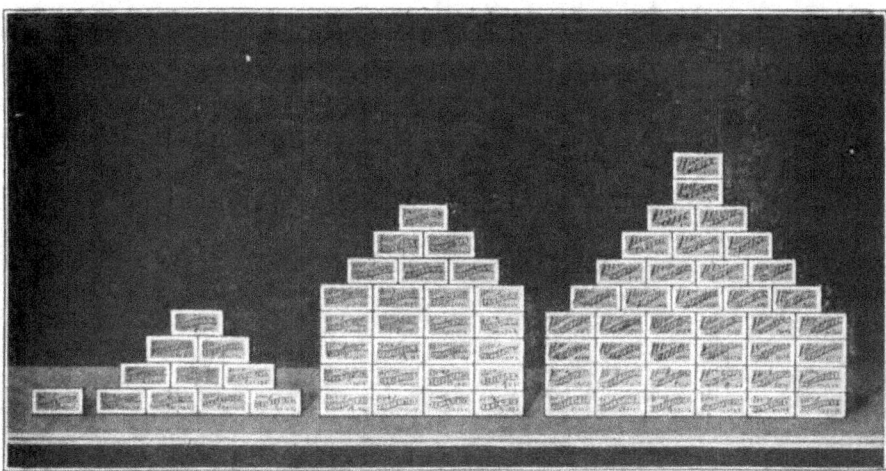

Courtesy Purdue Experiment Station

Hand Separator	Deep Setting	Shallow Pan	Water Dilution
Loss of Butter	Loss of Butter	Loss of Butter	Loss of Butter
1.2 lbs.	10.1 lbs.	26.2 lbs.	40.5 lbs.

When the cream separator is used on the farm, the milk can be separated immediately after it is milked and before it has time to absorb objectionable taints and flavors from outside sources. Cream that has been separated on the farm and properly cared for will make the very best grade of butter. With gravity and dilution methods of separation, it is impossible to determine the richness of the cream. It will be thin or thick, according to temperature, time of year, length of time the cows have been milking, and the breed of the cows milked. The per cent of butter fat in separator cream can be very closely regulated at all times, and a heavy or thick cream may be had as desired.

Government tests have proved that it costs from 10 to 15 cents per hundred pounds to haul milk the average distance of six miles to and from the creamery. This means a loss of from 2½ to 5 cents per pound on butter fat. Farmers who have been hauling milk to the creameries know that these figures tell only part of the story. Milk must be delivered every day. Very often it will mean many dollars to the farmer to have at work in the field the team which he must send to the creamery with the milk. Besides, there is the unpleasantness of having to go to the creamery day after day, through all kinds of weather, and over all kinds of road. In the spring of the year, when the roads are bad, the wear and tear on wagons and harness often amounts to as much as the milk is worth.

Cream can be delivered two or three times a week in summer, and in winter one trip a week is often enough. The expense of hauling cream can be reduced to very little by having one man do the hauling for a number of farmers.

FEEDING VALUE OF SEPARATOR SKIM MILK

Many careful tests have been made to determine the feeding value of skim milk, as it is now generally recognized that separator skim milk has many advantages over gravity or dilution skim milk for feeding calves and pigs.

Clean, sweet, separator skim milk is the best for dairy calves

The West Virginia Experiment Station has found that with eggs selling at 20 to 25 cents per dozen, skim milk used for moistening the mash fed to the chickens has a value of 2 cents per quart.

In "Feeds and Feeding," Professor W. A. Henry says: "The fat of milk has too high a market value with the dairyman to be used for calf feeding, and experience has shown that dairy stock of the highest quality can be produced from feeding skim milk."

Pigs thrive on separator skim milk

Skim milk separated on the farm immediately after milking will be clean, warm and sweet, and in the best condition for feeding. It is impossible to secure skim milk of this kind without the cream separator. Gravity skim milk will be cold and often sour before it can be fed. Even if it is sweet, to get the best results it must be warmed before feeding, which will cause trouble and a waste of time.

Dilution skim milk is so thin because of the water with which it is diluted that it has practically no feeding value, and calves will grow thin drinking it.

Regarding factory skim milk, Professor D. H. Otis says in a Bulletin issued by the Wisconsin Experiment Station: "Good calves can be raised on factory skim milk, provided the creamery is careful to receive only good sweet milk, so that the skim milk may be kept sweet until consumed by the calves. It should be borne in mind, however, that unless factory skim milk is heated sufficiently to destroy germ life, it is not only difficult to keep it sweet, but it may spread disease, especially tuberculosis, to the calves and hogs kept on the farm.

"It is much less work when the hand separator is used and the calves are assured of a more uniform feed. The calves are usually fed immediately after separating, while the milk is still warm and sweet. This uniformity of the ration and freedom from outside infection in milk is exceedingly important, and the hand separator deserves much credit for making this possible and practicable."

Professor W. A. Henry of the Wisconsin Agricultural College has said: "The dairyman who sells butter and feeds the skim milk to farm animals parts with an insignificant amount of fertility. When cheese is made, if the whey is returned to the farm a considerable proportion of mineral matter is conserved, but most of the nitrogen is lost. If whole milk is sold, the drain of fertilizing matter is considerable. These differences should always be borne in mind in conducting the various branches of dairy farming."

In a bulletin published by the Purdue Experiment Station we read, regarding separator skim milk, "It is perfectly fresh and sweet and can be fed to the calves while still warm. It is generally conceded that separator skim milk is worth about 25 cents per 100 pounds. The gravity skim milk is much older (12 to 30 hours), is cold and usually sour."

HOW THE CREAM SEPARATOR OPERATES

The separation of cream and skim milk by the action of the force of gravity is due to the difference in their specific gravities. It is also this same difference in specific gravity that makes it possible to bring about separation of cream and skim milk in the centrifugal cream separator. In fact, the action of centrifugal force is much the same as that of gravity, except that it works in a horizontal instead of vertical direction and in the I H C cream separator bowl is several thousand times stronger than gravity.

Centrifugal force is a force exerted outward from the center of the separator bowl and is produced by revolving the bowl at a high rate of speed. Just what the action of centrifugal force is can be best explained by a simple and often-used illustration.

When a ball attached to the end of a string is swung around in a circle, the ball, because of its weight, will exert an outward pull. The force exerted on the ball which makes it try to get away from the central point around which it is whirling is centrifugal force. When whole milk enters the separator bowl, it is acted upon by centrifugal force and the heavy milk solids are thrown to the outer wall of the bowl. The butter fat, which is the lightest part of the milk, is not so strongly affected and gathers near the center of the bowl where it is mixed with a small amount of skim milk and forms cream.

The amount of centrifugal force exerted outward from the center on the milk in a separator bowl is determined by the speed and diameter of the bowl. As the diameter of the bowl is decreased, the speed at which it is revolved must be increased or there will be loss of centrifugal force.

I H C Cream Separator bowls are equipped with an interior device composed of a tubular milk feeding shaft and a number of disks. The disks divide the milk into thin layers or sheets and centrifugal force acts upon each sheet of milk independently of the others. The disks increase the capacity of the bowl and reduce the speed at which it must be revolved by eliminating the necessity of forcing the skim milk solids through a thick wall of milk. This means greater durability because it reduces the strain upon the operating mechanism. When the heavy milk serum or skim milk is thrown to the outer edge of the bowl and the lighter fat globules gathered in the center, continuous separation is accomplished by the inflowing milk, forcing the already separated skim milk and cream up and out through their respective outlets in the top of the bowl. At the bowl outlets they are caught in skim milk and cream covers and conveyed to cans or other receptacles by means of spouts.

Healthy calves are found where separator skim milk is fed

THE BLUEBELL CREAM SEPARATOR

The Bluebell Cream Separator is a gear drive machine. Power is transmitted from the crank to the gears by means of a divided vertical steel shaft and the bowl revolved by a simple set of spiral gears. Briefly stated some of the most valuable features of this separator are:—

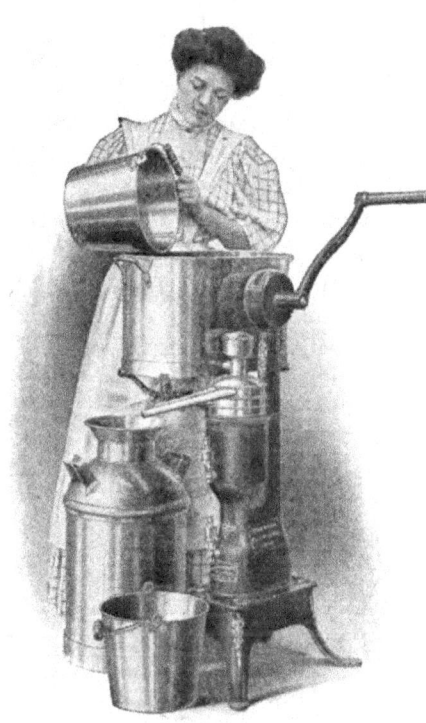

The Bluebell is a remarkably close skimmer because the bowl is built on the right principle and contains the best interior device ever manufactured.

The highest grade of steel has been used in the construction of the gears and spindles. This assurance of durability in the wearing parts is accentuated by phosphor bronze bushings.

It is light-running and its light-running qualities are not gained by light, cheap construction. It is friction that causes most separators to require a large amount of power. In the Bluebell, friction has been reduced to a minimum and consequently it is light-running. The gears are entirely protected from dust, grit, and milk, and at the same time are easily accessible.

The Bluebell has the strongest, simplest, and most effective top or bowl spindle bearing ever used in a separator. Instead of having a number of small, weak springs, this bearing has only one large, strong spring, which absorbs all vibrations due to starting and stopping.

The supply can is low and easily filled. The crank shaft is conveniently placed and does not force the operator into an awkward, uncomfortable position.

The bowl is equipped with a dirt arrester chamber, which removes all undissolved impurities before separation begins. As foreign substances are not spread all over the interior it is a simple matter to keep the bowl clean. This feature also insures a high grade of cream. The interior of the Bluebell bowl does not contain any of those forms of intricate construction which make the cleaning of other separators so difficult.

Bluebell Cream Separators are made in four sizes:

No. 1—Capacity, 350 lbs. of milk per hour
No. 2—   "   450 "   "   "   "
No. 3—   "   650 "   "   "   "
No. 4—   "   850 "   "   "   "

THE DAIRYMAID CREAM SEPARATOR

One of the features of the Dairymaid Cream Separator that has proved remarkably efficient, is the method of transmitting power from the crank to the gears. Power is transmitted to the gears by means of a drive chain which operates over a large sprocket on the crank shaft and

a small sprocket on the intermediate gear shaft. The chain drive for separators has been in use in Europe for many years, but was first used in this country in the Dairymaid. The fact that many attempts have been made to copy the chain drive principle of the Dairymaid, is good evidence of the desirability of a machine so equipped. The advantages of the chain drive are, that the wear on the gears is considerably reduced, the machine runs smoothly and noiselessly and, because of the difference in the size of the sprockets, the crank turns easily.

Another distinguishing feature of this separator is that it has a double oiling system. The gears and lower bearings are oiled by an oil bath and the neck bearing is oiled by a sight feed oiler located on the front of the bowl housing. Oil from this oiler flows down through the neck bearing, lubricating it and renewing the supply of oil in the oil bath.

The bowl spindle or neck bearing of this separator is very simple and practically trouble proof.

Other features of this separator which are similar to those of the Bluebell Cream Separator are, the interior device in the bowl, including the dirt arrester chamber, the easy cleaning qualities of the bowl due to the absence of intricate forms of construction, the use of phosphor bronze bushings, low supply can, and conveniently located crank.

Dairymaid Cream Separators are made in four sizes:

No. 1—Capacity, 350 lbs. of milk per hour
No. 2— " 450 " " "
No. 3— " 650 " " "
No. 4— " 850 " " "

THE LILY CREAM SEPARATOR

A few of the features which contribute toward making the Lily a separator that will prove one of the best investments a farmer can make in this line are:

The Lily bowl has an interior device composed of a number of disks and a milk-feeding shaft. The milk-feeding shaft used in this bowl is a decided improvement over any that has been used in other separators during past years. The milk-feeding shaft of the Lily is so designed that the entire skimming surface of the disks is utilized, whereas in the old type, where the milk is fed through openings in the ends of the wings nearly one-half of the skimming surface of the disks is lost. The advantages of a disk bowl have been so often demonstrated that every person who has given separator construction even a very little consideration knows that this is the only correct principle upon which to build a separator bowl.

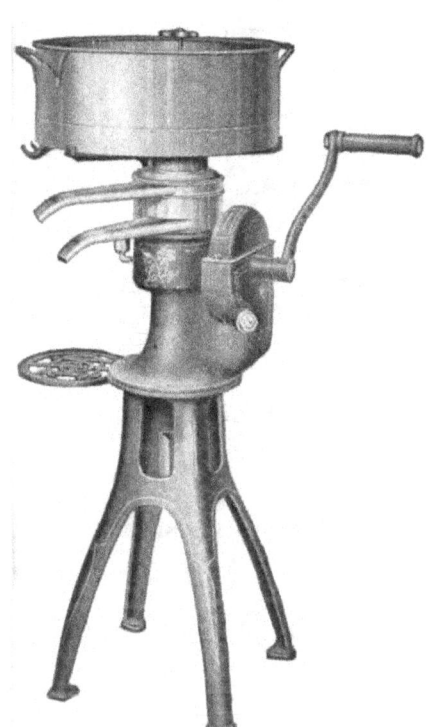

The Lily cream separator is not excessively heavy nor clumsy, yet every part has more than necessary strength and wearing quality to withstand years of continuous use.

One of the most desirable features of a cream separator is simplicity. A separator that is simple in design will not get out of order easily nor require numerous adjustments. An examination of the Lily will show that it is simple and has very few parts.

The Lily is an easy, smooth-running separator, and very little effort is required on the part of the operator to turn the crank.

Every part of the interior of the Lily bowl presents a plain, smooth surface to which dirt and milk do not adhere, and consequently it is an easy matter to wash this bowl and keep it thoroughly clean.

No cream is permitted to collect in the tubular shaft, which makes this part also easy to clean.

The oiling facilities provide for a thorough lubrication of all moving parts. The gears run in a bath of oil, and even the neck bearing is lubricated by the oil bath. This forms the most practical and reliable method of lubrication ever used on a separator.

The supply can and crank are conveniently located for easy filling and turning. The Lily Cream Separator is made in four sizes:

No. 1—Capacity, 350 lbs. of milk per hour.
No. 2—   "   450 "   "   "
No. 3—   "   650 "   "   "
No. 4—   "   850 "   "   "

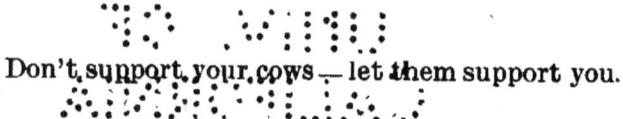

Don't support your cows — let them support you.

The farmer who reads does not have to look for profits with a microscope.

Success in farming is only attainable through study and application of scientific principles.

Anybody can milk cows, but it takes a man with brains to milk the right sized profit out of them.

No man knows all there is to be known about farming—let us all get together and learn from each other.

The amount of brains you put into your work determines the amount of pleasure and profit you will get out of it.

Agricultural progress has been made by men who were not satisfied with what was good enough for their grandfathers.

Don't keep three cows to produce 12,000 pounds of milk when two better cows will do it with the same amount of feed.

There is no branch of agriculture that takes as little fertility from the soil and at the same time returns as good a profit for the farmer as dairy farming.

The man who learns to get two pounds of butter from the same amount of feed that before produced only one, is going to get from under the mortgage quick.

Wherever the farm products have been turned into butter for a number of years, there has been a steady increase in the crop producing capacity of the soil.

The successful man in any business is the one who can and will make use of the experience of others — who has the courage to discard his own errors and adopt the truths discovered by others.

A man who would annually sell a few acres of his farm instead of cultivating it would be considered a very poor farmer. Yet, this is just what is being done when crops which take a large amount of fertility from the soil are sold off the farm.

www.ingramcontent.com/pod-product-compliance
Lightning Source LLC
Chambersburg PA
CBHW062228220526
45471CB00009B/3393